Entrepreneurship in ACTION

A RETAIL STORE SIMULATION

Entrepreneurship in ACTION

A RETAIL STORE SIMULATION

ROSALIE J. REGNI
Virginia Commonwealth University

JIMMIE G. ANDERSON
U.S. Small Business Administration (retired)

Fairchild Books, Inc.
New York

Director of Sales and Acquisitions: Dana Meltzer-Berkowitz

Executive Editor: Olga T. Kontzias

Senior Development Editor: Jennifer Crane

Development Editor: Joseph Miranda

Assistant Development Editor: Blake Royer

Art Director: Adam B. Bohannon

Associate Art Director: Erin Fitzsimmons

Production Director: Ginger Hillman

Senior Production Editor: Elizabeth Marotta

Copy Editor: Joanne Slike

Interior Design: Julia Nourok

Cover Design: Adam B. Bohannon

Library of Congress Catalog Card Number: 2007943510

ISBN: 978-1-56367-595-9

GST R 133004424

Printed in the United States of America

TP 09

Contents

Extended Contents

Preface

The Butcher, the Baker, the Candlestick Maker

Going into business takes a leap of faith. Who hasn't at one time or another thought about what it would be like to own their own business, be their own boss, and just sit back and rake in the dollars? How many times have you heard someone say something like this: "Wow! You are a great cook! You should open a restaurant." To make that dream a reality and actually take that leap requires a lot of thought, hard work, ingenuity, planning, help, sometimes luck, and in almost all cases money.

This book will help the reader delve into those myriad nuances of **entrepreneurship** as well as provide a tangible simulation of the processes necessary to start a business. The information, CD-ROM, and resources provided are beneficial for the student who is completing course work or the person who is thinking of opening his or her own business. This book is also useful for those already in business who may be looking for guidance or a resource tool.

The CD-ROM provides video interviews with real store owners; financial documents; and examples of advertising material, floor plans, and stationery from Rose Knows Clothes—the simulated store in this book. The CD icon appears throughout the book and alerts readers to material that can be found on the CD-ROM.

Helpful tools throughout the text assist readers with making the decision of not only whether to go into business but also how to go about it. In addition, a simulation of a business start-up is included, which will give the reader hands-on experience. The following questions will also be discussed and answered:

1. Do I want to start my own business?
2. Am I suited to own my own business?

3. Where can I go to get help?
4. What resources are available to me?
5. How much money do I need?
6. Where do I go for money?
7. What kind of business do I want?
8. What are my qualifications?
9. What part does government have in my business?
10. What makes up a business plan?

Going into business is not for the faint of heart. It takes lots of guts and determination. Once you own your own business, the days of starting at 9:00 A.M. and going home at 5:00 P.M. are pretty much a thing of the past. Plus, when you worked for someone else, you were only responsible for one job.

As a business owner, you will now be required to know how to do all the necessary jobs and most likely you will have to do all the work—at least in the beginning. In the nursery rhyme "The Butcher, the Baker, and the Candlestick Maker," there were three different people. In the world of business ownership all of those jobs are done by one person—you!

Acknowledgments

To prepare a textbook simulation of this magnitude, input from many industry professionals is crucial to the accuracy and validity of the material. We owe thanks to many such professionals who have very graciously given us their time, expertise, and insight into their vast stores of experiences and knowledge of entrepreneurship.

First, a very special thanks to Michael A. Sisti, branding, marketing, and advertising consultant; he wrote a significant part of Step 7, "Marketing Plan," and created the artwork for that step and the marketing section of the CD-ROM. Much appreciation is owed to Art Collins, retired Director for Government Contracting, Small Business Administration; he took on the daunting task of writing the very intricate and detailed material for Step 11, "Financial Documents for Evaluation and Tax Preparation."

Gratitude is also owed to the following individuals who made significant contributions to our text: Leroy A. Keller, Lender Relations Specialist for the Richmond, Virginia, Small Business Administration, for his initial input and review of our proposal; Diedra Arrington, Adjunct Professor of Fashion Merchandising at Virginia Commonwealth University and former Vice President, Peebles department store, for her input on and review of Step 6, "Merchandising Your Store"; Laura Golluscio, Human Resources Manager, First Environment, for her help with Step 8, "Personnel and Sales Management Plan"; Thalhimer Commercial Real Estate and especially agents James Ashby and Eric Robison for providing us with information on and use of our simulated store location in Carytown; Carol Amernick, accountant, who provided much needed help with details on costs and procedures; J.R. Downey, real estate agent with Alfred J. Dickinson, Inc.; John Nelson and Jason Faulkner of VAVS Productions, who recorded and produced the video and photographs for our store owner interviews; Karen Videtic, Chairperson, Department of

Fashion, Virginia Commonwealth University, for several good ideas that she gave us for our book and CD-ROM; Jennifer Hamilton, Assistant Professor, Department of Interior Design, for the designs and drawings of our simulated store; Krutika Kotral, VCU Student, for design of our store Rose logo; Jeffrey D. Virgin, Insurance Agent at Straus, Itzkowitz, & LeCompte; Spencer Lauterbach of Isis Business and Accounting Software Consulting.

A very special thanks also to the Richmond small store owners who took the time to allow us to interview them in their stores and tape those interviews for the CD-ROM: Heidi Story of Heidi Story, Lisa McSherry of Lex's of Carytown, Heather Teachey-Lindquist of Que Bella, Libby Sykes and Deborah Boschen of Pink, and Casey Longyear and Marshe Wyche of Rumors.

There are many others who have provided input and guidance, and we would like to thank all of them, including the reviewers who were selected by the publisher to provide guidance in this endeavor.

We certainly could not end this list of thanks without expressing our appreciation to our spouses (and friends), Patricia Wiseman Anderson and Albert Regni, for their support and encouragement during the many months of creating this simulation. We are also extremely grateful for the patience, input, encouragement, and assistance of Joseph Miranda, our Development Editor, and Olga Kontzias, Executive Editor, and all of the great team at Fairchild Books.

INTRODUCTION
Getting Started

PROJECT OBJECTIVES

This step in the simulation will guide you through:

- Objectives of the simulation
- How the simulation works
- Evaluating potential as an entrepreneur
- Where to start and resources for help
- Definition of a business plan
- Preparing an action plan

Objectives of the Simulation

Webster's dictionary defines a simulation as "an act of pretense." The simulation presented will be just that—a pretend effort at opening a business. All aspects of the process will use real-world applications and the reader, whether a student or potential entrepreneur, will be able to take the applications and apply them to his or her own study process or business situation.

By working with and using the simulation, the reader will get a better understanding of the processes involved in business ownership. The simulation takes the reader from the start-up decision through all the steps required—finding a location, developing the business plan, opening the business, financing the business, marketing the business, as well as running and maintaining the business. Upon completing the simulation, the reader will have a better understanding of ***entrepreneurship*** and how it works.

How the Simulation Works

Each subsequent chapter in this text takes you through a step, which will form one of the building blocks in creating the business plan. You will be given key information to assist you in the development of your plan, but it is your responsibility to do additional research to gather all the data needed to complete the simulation. The CD-ROM contains blank forms to be used in the development of the business plan, and the appendix has a completed plan for a women's clothing boutique. In this chapter, you will find some helpful resources to get you started in your research. A list of additional reference tools is included on the CD-ROM and in the appendix.

To illustrate the various parts of the business plan, we will use examples of a plan, which has been developed for a fictitious women's clothing store in Richmond, Virginia, in a well-known shopping area called Carytown. The information in the plan is readily applicable or transferable to a similar small store in any community in the United States. It will be your responsibility, as part of your research, to find similar information for a store in your particular city.

If you are an entrepreneur using this text as a guide, you probably already have a city and perhaps even a location in mind. If you are a student, you should find an empty small store in the town where your school is located. This will help the simulation seem more real to you and actually make the task more doable. This step will help get you started on where to obtain the necessary information, and the CD-ROM list of resources will give you additional places to search.

Evaluating Your Potential as an Entrepreneur

Everyone—or at least everyone we know—seems to think that running a business is a snap. You get a product or a service that everyone wants and you sell it to them at a reasonable price. Then you just sit back and count your money. While some would laugh at that pie-in-the-sky scenario, deep down many have probably wondered at one time or another whether or not they had what it takes to own and run a business.

In fact, the very first questions that anyone who wants to start a business should ask him- or herself are these: Am I cut out for this? Do I know what I am doing? Am I prepared to fail? Am I prepared to succeed? How is this going to affect my family? Do I have their support? What are the odds of my success or failure? What are the consequences of my success? What are the consequences if I fail?

In my more than 25 years with the ***U.S. Small Business Administration (SBA)***, I've met hundreds, if not thousands, of entrepreneurs and potential entrepreneurs. Those who succeeded had a few common characteristics: they were very determined, they were very focused on their vision, they were very smart, and they were lucky. Those who did not succeed lacked at least one of those traits.

Many will dismiss the last trait—luck—and say that you make your own luck. While that is true to a certain extent, just know that a lot of intangibles come into play in starting a business, as they do in everyday life. Know that curveballs will be thrown at you. Just do your best to swing at them and realize that sometimes you will strike out but other times you will hit a home run. Those that can make the best lemonade from the lemons that life throws at them generally are the ones that succeed.

Before deciding whether or not to start your own business, you must first perform a self-assessment of your entrepreneurial skills. You can find one assessment at http://www.gsu.edu/~wwwsbp/entrepre.htm, and many others are available online. The SBA also offers an entrepreneurial assessment on its Web site at www.sba.gov.

Where to Start

Once you decide to try your hand at entrepreneurship, you will find that many resources are available to help you and weeding through them can be a daunting task. For instance, if you enter "how to start a business" into a popular online search engine, you'll find that tens of millions of sites are referenced. It would take a lifetime to research all of those sites to glean the pertinent information.

Know from the beginning that all levels of government are interested in helping you succeed. Successful small businesses are the engine that runs the economy of the United States. In fact, according to the SBA, the following are 10 reasons why we should love small businesses:

1. The latest figures show that small business creates 65 percent or more of America's net new jobs.
2. There are approximately 4,115,900 minority-owned businesses and 6,492,795 women-owned businesses in the United States, and almost all of them are small businesses.
3. The United States saw an estimated 580,865 new small firms with employees start up in the last year measured.
4. Small businesses are 97 percent of America's exporters and produce 26 percent of all export value.
5. Home-based businesses account for 53 percent of all small businesses.
6. Small businesses employ 50.1 percent of the United States' non-farm private-sector workers.
7. The more than 24 million small businesses in the United States are located in every community and neighborhood.
8. Small patenting firms produce 13 to 14 times more patents per employee than large patenting firms.

9. Small businesses create more than 50 percent of the American non-farm private gross domestic product (GDP).
10. Small businesses make up 99.7 percent of all United States employers.

Source: Office of Advocacy, U.S. Small Business Administration, News Release SBA 06-04 ADVO.

The most comprehensive federal government resource for the budding entrepreneur is the SBA. According to its Web site, "The SBA helps Americans start, build and grow businesses. Through an extensive network of field offices and partnerships with public and private organizations, SBA delivers its services to people throughout the United States, Puerto Rico, the U.S. Virgin Islands and Guam."

Each state has its own business development agency that aids citizens in starting businesses. While each may have a different title, the function is basically the same. For example, the Commonwealth of Virginia has its Department of Business Assistance (DBA). According to the DBA's Web site (http://www.dba.virginia.gov), the agency "is the economic development agency devoted to the growth and success of the Commonwealth's business community. Established by the Virginia General Assembly in July 1996, the DBA rounds out our state's economic development program by ensuring that businesses not only find Virginia an excellent place to locate, but also an ideal place to grow, expand, and make additional investments." The resources at the end of the book list contact information for the business development agencies.

Many local governments, both city and county, may also have their own business development offices, and these organizations are particularly helpful because their information will be geared to the locale. Employees in these offices will have up-to-date information on local business license requirements, location information, business climate, tax rates, and so on. An example of a local government business development office is in Henrico County, Virginia. Their Web site (http://www.henrico.com/assistance_program.php)

offers help in such areas as business plan development, business facilities lease/purchase, business finance/banking, marketing, advertising, insurance, and many others.

Other available resources are libraries, universities, chambers of commerce, merchants associations, trade associations, business development centers, and so on. The list of available help is almost endless, and the research experience can be mind-numbing. When embarking on this experience, persistence and perseverance are definitely virtues.

The Business Plan

As the SBA says, "A written guide to starting and running your business successfully is essential. This plan will encourage loans, promote growth, and provide a map for you to follow."

Everyone at one time or another has heard that the three most important things for a business are location, location, location. While location obviously is crucial to the success of a business, there are in fact three other elements that are even more important: planning, planning, planning. Try going into a bank without having some semblance of a business plan, and they will laugh you right out the door.

If you have any hope of having your business succeed, not only do you need to plan every detail of every step, but you will also need to update your planning as you move along the path of entrepreneurship. Planning never ends; it is always evolving. Most businesses fail because they lack proper record keeping and, often, behind every poor record keeper you will find a poor planner.

Whoever said that "failure to plan on your part does not create an emergency on mine" knew what he or she was talking about. You will rapidly find, if you do not know it already, that no one is as interested in your business as you are—except for maybe the people to whom you owe money. Your deadlines generally only mean something to you, and you cannot depend on others.

You can only depend upon yourself. My father-in-law, who was a very good carpenter and very wise in the ways of getting things done right the first time, had a favorite saying that was appropriate not only for carpentry but also to every aspect of planning: Measure twice, cut once.

Types of businesses are generally categorized as follows:

1. Manufacturing
2. Nonmanufacturing (including retail stores)
3. Construction
4. Service
5. Professional service

All business plans have basic characteristics, but they also must be modified depending on the type of business involved. A plan for a retail business, which falls under the nonmanufacturing category, will need to address ***inventory*** but will have little need for heavy equipment. Conversely, a plan for a construction business will need to address heavy equipment but will have little need for inventory planning.

The SBA's model for a business plan is well tested, and so we use its basic elements here. The body of a business plan can be divided into four distinct sections:

1. Description of the business
2. Marketing
3. Finances
4. Management

The agenda should include an executive summary, supporting documents, and financial projections. Although there is no single formula for developing a business plan, some elements are common to all of them, as summarized in the following outline. This will serve as the outline that you will use in completing the simulation. Your final plan will be organized in this order, and the balance of the steps will guide you through each part of the outline.

Keep in mind that we will not work through the simulation of the plan in exactly the order presented, but you should use the outline to organize your final project. We believe that it is better for students, in particular, to first gather the information in the order described in each step and then wait until the end to complete Sections I, II, III, and IV. The simulation instructions at the end of each step guide you through the steps needed to complete the simulation, and the outline allows you to properly organize the information for the final completed plan:

Elements of a Business Plan

I. Cover Sheet
II. Statement of Purpose (also called Summary Page)
III. Table of Contents
IV. Description of the business
V. Product and Service Plan with appendix items
    A. Description of product, product assortment plan, vendors, and key elements of services offered
    B. Appendix items
        1. Opening assortment plan (Step 6)
        2. List of key vendors with addresses, products, and key price points
VI. Location Details with appendix items
    A. Description of location plan with benefits and challenges of store site selected. It also details the look and visual merchandising plan for the exterior and interior of the store, describing how these elements will contribute to the overall image and success of the business.
    B. Appendix items
        1. Map of the area where your store is to be located
        2. Lease agreement with Store Location Fact List (Table 2.1)

3. Floor plan of store showing location of fixtures and merchandise (similar to Figure 9.5)
4. Store setup budget (Table 9.2)

VII. Marketing Plan with appendix items

A. Customer Section—Describes the target customer with a demographic and psychographic profile

B. Competition—Analyzes direct and indirect competitors and addresses how the new store will deal with competitors

C. Pricing and Sales—Describes store policies for setting retail prices (both regular and promotional) and details plans for achieving mark-up and margin goals

D. Advertising/Promotions—Details store marketing program with budgets for marketing the store image and products and for attracting target customers.

E. Appendix Items
1. Calendar of promotional/marketing events for first year of business (Table 7.1)
2. Tabulation of marketing costs for both pre-opening and ongoing (also on Table 7.1)

VIII. Personnel and Management plan with appendix items

A. Describes qualifications of owners and management and outlines plans for hiring, training, and rewarding sales associates for the store

B. Appendix items
1. Sample weekly personnel coverage plan with salary costs as compared to projected weekly sales plan
2. Organizational chart for the business
3. Sample job description for each different position
4. Resumes for each partner
5. Personal Financial Statements for each partner

IX. Financial Data

   A. Loan applications (for entrepreneurs only)
   B. Sales Projections by year (Table 3.2) and month (Table 5.5)
   C. Gross Margin Plan (Table 5.1)
   D. List of fixed expenses (Table 3.2)
   E. Breakeven analysis (Table 5.7)
   F. Pro-forma profit and loss statements (Table 5.9)
   G. Capital equipment and supply list – Capital Spending Plan (Table 10.1)
   H. Pro-forma cash flow (Table 10.2)
   I. Pro-forma balance sheets (Tables 11.1 and 11.3)

      Note: For each financial document, you will need to provide the following:

      1. Three-year summary
         a. Detail by month, first year
         b. Detail by quarters, second and third years
      2. Assumptions upon which projections were based (also called justifications)

         Note: Students will not be required to complete three years of *all* financial documents. See simulation instructions at the end of each step.

   J. Additional Supporting Documents (these will not be required for student projects)
      1. Tax returns of principals for the previous three years
      2. For franchised businesses, a copy of franchise contract and all supporting documents provided by the franchiser
      3. Copy of licenses and other legal documents
      4. Copies of letters of intent from suppliers

X. Bibliography—Details sources for your research, interviews with business professionals, and contacts for financial information

Action Plan

The idea has been formulated. The resources have been identified. The type of business plan has been formatted. So, what happens next?

The first objective is to complete the business plan using the resources that are available. Once the business plan process is under way, you will find that as you answer one set of questions, another set pops up. You must be honest with yourself about the goals that you are setting and your means of obtaining them.

The hard reality of the world is that most businesses fail. Most fail because of poor planning. It cannot be stressed enough that *realism is the key*. Know what you want to do and be realistic about your means of achieving it. If you have limited financial resources and limited means of obtaining them, then your planning needs to reflect it. One of the biggest mistakes committed by many newly minted business owners is that they have unrealistic expectations. The adage that "those who fail to plan are planning to fail" is more true than ever when it is applied to starting a small business.

Simulation

1. Complete a self-evaluation exercise, such as the one referred to earlier in this step. (Students: Your professor can also provide you with an evaluation.)
2. Using the results of the self-evaluation, make a list of your strengths and weaknesses as they relate to running a business. You will use this list later when you complete the personnel plan. At that time, you will be asked to write about your background, professional experiences, strengths and weaknesses, and how these will contribute to or potentially hinder your success. You will also be asked to discuss your plans for overcoming or compensating for any weaknesses (e.g., lack of experience in managing a store).

STEP 1
Choosing a Business and Product Line

"Accept the challenges, so you may feel the exhilaration of victory." —*General George S. Patton*

PROJECT OBJECTIVES

This step in the simulation will guide you through:

- Selecting the product line and categories for your store
- Identifying the niche that will separate your store from the competition
- Conducting industry research to ensure your expertise in the product arena
- Creating a regional market feasibility study to assess your chances of success in your selected area
- Building a profile of the customers in your target market
- Establishing your business structure: Do you need a partner? If so, what kind?

The simulation instructions at the end of this step require you to identify your product line for your store, identify your ***niche***, and build a profile of your ***target customer***. The information in the step will assist you in that effort, and will also provide guidance and information for some of the other steps in building your business plan. As mentioned in the Introduction, some parts of the business plan are located chronologically at the beginning of the finished plan, but we will write about them toward the end of the process, as these parts act as summaries of the entire plan. This step covers some of the information included in those sections, including a summary page, or ***letter of intent***, and business description. As you work your way through the plan, start thinking about the necessary information for those parts of the plan.

Selecting the Product Line and Categories for Your Store

The first and most important rule of selecting a product line for your store is to select one that you feel passionate about—one you are willing to spend time, money, and energy learning about and developing into a successful business. Because most of us will spend more time working than doing anything else, your work should be a part of your life that brings you fulfillment and joy. This is especially true with a new business venture, which is really a commitment to "working 24/7."

Of nearly equal importance is gaining prior experience with the product category you select. Before you embark on a business venture that sells women's clothing, for example, it is best to spend some time working in a store that sells women's clothing so that you can learn techniques and procedures that work and (of equal importance) those that don't. At the same time that you are working to gain this experience, you should be researching the market and learning everything that you can about the women's clothing business.

Your research and experience in the business can help you to take the next step: selecting categories for your product line. If you use a traditional department store as a starting point, look at the departments within a typical store. Each of those departments should be able to stand alone as a specialty store. Within the department (women's career wear, for example), you might find suits, separates (blazers, skirts, and trousers), blouses and sweaters, outerwear, dresses, and perhaps others. Each of these is labeled a category. The combination of these categories makes up the department, and each could stand alone as a specialty store. As a small store owner, however, you may not be able to carry all of the categories illustrated in the example above. You would, in fact, make an error of judgment if you tried to carry all of these in a store that is too small to accommodate them. It would be better for the small store owner to select the three to five most important categories so as to allow flexibility and sufficient inventory dollars and offer a broader selection within each category. It is difficult for an entrepreneur to consciously walk away from a potential sale, but the entrepreneur must always keep in mind that the small store just doesn't have the luxury to be all things to all people.

In our women's career wear boutique, when researching category options, the entrepreneur might use deciding factors such as assortments of local competitors, perceived demand by target customers, trends in the women's clothing business, and knowledge of success or failure of these categories among both local and national retailers. Table 1.1 provides examples of categories for selected types of small fashion boutiques. You will note that not all product lines or categories are included in the table. These are merely examples to help you hone in on your own categories and product lines. One can also see that some of the categories can be expanded into separate specialty stores. Victoria's Secret has proven that there is a strong market for a lingerie and intimate apparel store.

Selecting product lines and categories is crucial to the beginning stages of a business plan, but it is not enough. Even more important is the determination of your store niche.

TABLE 1.1 SAMPLE CATEGORIES FOR SMALL RETAIL BOUTIQUES					
The wise retailer will focus on three to five of these categories (depending upon store physical size) and not attempt to offer every category within a product/customer designation.					
WOMEN'S CAREER WEAR	**JUNIOR'S CASUAL WEAR**	**MEN'S WEAR**	**YOUNG MEN'S FASHION**	**CHILDREN'S WEAR**	**HOME FASHIONS**
Suits	Bottoms	Suits	Bottoms	Boy's bottoms	Bath
Dresses	Jeans	Sport coats	Jeans	Boy's tops	Kitchen
Lingerie	Dresses	Dress slacks	Shirts	Girl's bottoms	Lamps
Tops: shirts, sweaters, blouses	Activewear: yoga, exercise, swimwear	Dress shirts	Sweaters	Girl's tops	Small furniture: tables, bookcases
Separates: skirts, pants, blazers	Tops: blouses, camisoles, sweaters	Underwear	Casual jackets	Overalls	Decorative: vases, mirrors, trays
Shoes	Shoes	Shoes	Shoes	Jackets and coats	Window treatments
Accessories: jewelry, hosiery, hats, handbags	Accessories: jewelry, socks, tote bags, belts	Accessories: ties, belts, handkerchiefs	Accessories: hats, belts, jewelry	Infant's wear	Bedroom: linens, curtains, coverlets
Outerwear	Coats and jackets	Outerwear and tuxedos	Swimwear	Toys, gifts, accessories	Gift items

Identifying the Niche That Distinguishes Your Store from the Competition

Webster's dictionary defines niche as "a place or position suitable or appropriate for a person or thing: *to find one's niche in the business world.*" There are several important points in this definition. First, the business (thing) and the owner (person) become one. The new business will be a direct reflection of the owner and his or her vision, so it is crucial that the owner find ways to entice and excite the potential target customer to shop in the new store. The difficult question to address is "Why would a customer spend money in my store rather than at my competitor's down the street or across town?" In their excitement about the new venture and their own ideas, many entrepreneurs forget that others might not be as immediately excited or understand the vision as well as the one who conceived it. Consumers, typically, are both skeptics and creatures of habit. You need a niche for your store in order to entice them to visit and to purchase. Sharon Brooks, owner of SB&A advertising agency in Richmond, Virginia, calls this "offering the customer a benefit with a ***brand promise.***" According to Ms. Brooks, once you have established the benefit that brings in the consumer, you must always deliver on the brand promise that your store represents.

The second element of the definition is "suitable or appropriate." Before opening just another women's clothing store in town, the entrepreneur must perform a thorough market research study on the competition. Several basic questions must be answered:

1. Is there room for another women's clothing store [for example] in my chosen town and area?
2. What do the competitors offer in terms of service, selection, and price?
3. What is the niche of each of my competitors? What do they do that is uniquely theirs?
4. What can I do to make my store different or unique? Can I offer a different selection (product that is missing from my market)? Can I provide services that the consumer wants but that are not currently offered by my competitors? Is there a price level that is desired but not offered to local consumers, yet is also one that I can realistically deliver?

Heidi Story, owner of Heidi Story, her self-titled fashionable women's boutique in Richmond's trendy Carytown shopping district, moved her store to Richmond, Virginia, from Brooklyn, New York, about two years ago. She recognized that she would need a niche to separate her store from the many other women's boutiques in Carytown. So, in addition to careful attention to the differentiation of her assortment, Ms. Story decided to use one room of her new location to set up a workroom. In that workroom, she teaches sewing classes to both children and adults. She also does alterations there and works with customers who desire custom-made garments designed by Ms. Story. In this way, she has distinguished her store and created a niche that makes her store different from others in the shopping area. Figure 1.1 shows the workroom/classroom in the Heidi Story store.

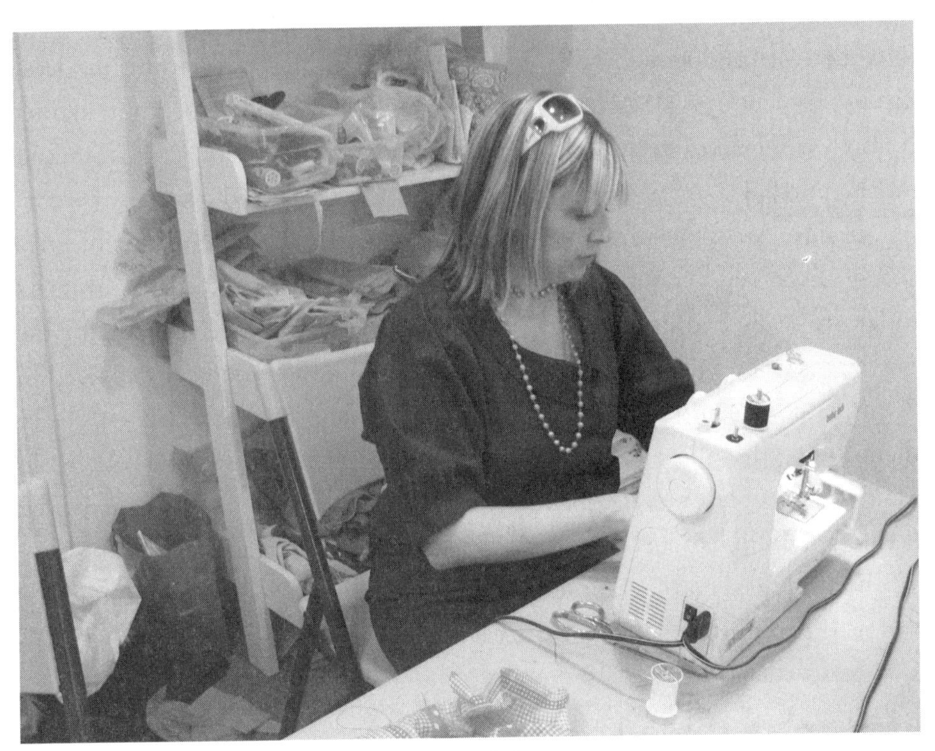

Figure 1.1 *Sewing/alteration room in Heidi Story store.*

We discuss the analysis of the competition more thoroughly in Step 7, on marketing, and there you will prepare a complete profile of your key competitors and a ***market condition report***. For now, however, you should not proceed until you are comfortable that your store will fulfill an existing need or offer a category of product currently not available in your market—in other words, until you have identified your store niche.

Another consideration in identifying your niche is to understand that, as a small store, you will not be able to satisfy the needs of all of the consumers in your town or market. You will need to create a profile of your target customer (discussed later in this step) and aim all your efforts toward enticing and satisfying that group of consumers. According to the SBA, you should evaluate the results of your market research or feasibility study to determine "the right configuration of products, services, quality, and price that will ensure the least direct competition" ("Starting Your Business," www.sba.gov). Your market research will help to determine what factors motivate your target customer to make a purchase. Are people in your selected area motivated primarily by current trends? Are they more likely to respond to the right combination of price and quality? You should also keep uppermost in your mind that consumers today are savvy and demanding and that they expect to receive value for the money that they spend; value meaning a perception that you have delivered to them the combination of price, quality, service, and selection that will leave them feeling that their money has been well spent. It is also good to keep in mind that people of every income level and every spending category, even luxury goods, like to feel that they have discovered a bargain.

With all these considerations in mind, you are ready to begin your industry and market research and to formalize a market feasibility study.

Industry and Market Research

The savvy entrepreneur will realize that all research and business plans should focus on "the customer and what he needs and wants to buy, rather than the product and service you want to sell" (Kuebler, *Building Your Future in Self Employment*, p. 33). This section concentrates on (1) what kind of research to pursue and (2) where to conduct the research. In the Introduction, we talked about resources to help you develop your plan, and the CD-ROM has additional sources and examples of research. These will give you some key places to begin your research, but you may need other sources as well, depending upon your location and type of business.

In general, your research should focus on the following:

- Determine the local need for your product and/or service.
- Explore the industry and market for your products, both nationally and locally, to hone your expertise in your product lines. This step is very important, as many of the factors that affect large companies will also impact small businesses in the same product lines.
- Evaluate the strengths and weaknesses of your direct and indirect competitors.
- Build a profile of your target customer(s).
- Select the best location for your business (this is covered further in Step 2).
- Set retail prices for each category of your product assortment.
- Assemble a list of the best-performing suppliers for your product categories.
- Analyze marketing and promotional efforts to determine what will work best for your store (discussed further in Step 7).

These areas will provide the structure for your initial research. Research will continue throughout the formation of your business plan in the areas of fixed expenses, necessary capital investment, sales planning, assortment building, sources of financial support, as well as other areas that we will discuss as the plans unfold.

One important tool that has commonly been used for the last few years to determine the potential success of a business idea is the SWOT analysis, for Strengths, Weaknesses, Opportunities, and Threats. The analysis helps the entrepreneur think through the most important considerations of a potential new business, and may assist in determining whether to proceed, stop, or revise the direction of the endeavor. Since this concept is taught in many business-related classes, we will only mention it here; details, however, plus an example of a SWOT analysis can be found on the CD-ROM.

Identifying a Target Market

The best way to start the process of identifying and locating your target customers is to develop a profile of consumers who live in and near the area where your store will be located. We discuss selecting a store location in Step 2, but it is a good idea to first understand how to identify customers before you settle on a specific site. When considering how customers will typically shop for your products, ask the following questions:

- Will they get in the car and make a special drive to find your products? Is your store a "***destination***" store?
- Will they more likely purchase your merchandise if there are other stores in the immediate area (strip mall, for example) that carry products they want or need?
- Will most purchases result from impulse buys made while customers are browsing in a particular area (downtown shopping district, for example)?

If most purchases will come from customers who live or work in the immediate area, you can start by collecting data about the ***census tract*** immediately surrounding your store. By consulting http://factfinder.census.gov/, you can gather population, housing, economic, and geographic data about the people who live and work in the census tract(s) where you are targeting your customer base; this information is labeled ***demographic***. In the simulation at the end of this chapter, you will be asked to gather data about your customers that involves details about the information available to you through the census bureau. You will also be asked for information about your locality. You can find information for your local and state demographics and economic statistics at sites such as http://usa.gov/agencies/state_and_territories.shtml and county government information at http://www.naco.org/. Most states, cities, towns, and counties have their own Web sites, and these can provide very useful information to help you build the target customer profile as well. Your library may have resource books such as *Community Sourcebook of County Demographics* and *Community Sourcebook of Zip Code Demographics*, where you can search for information on demographic, geographic, lifestyle, business, and consumer expenditure data. Additional sources are listed at the end of the book.

After collecting and evaluating demographic data on your potential customer base, you should attempt to find some ***psychographic*** data. This is more difficult to find, as it is not data that can be surveyed and enumerated like the demographic data of incomes, ages, and number of children, for example. Psychographics refer to such considerations as lifestyles, values, motivators, interests, and cultural influences, and explores the primary motivation for an individual's purchasing behaviors. They are ***qualitative*** instead of ***quantitative***. There are sources, however, for this information as well. To get a feel for psychographics, take the VALS™ (Values, Attitudes and Lifestyles Survey) test at www.sric-bi.com/VALS/. There is no charge to take this test. The VALS survey was developed and is owned by SRI Consulting Business Intelligence. It attempts to explain and predict adult consumer behavior by the use of an eight-segment framework that identifies primary motivations and resources to show why a consumer buys some things and not others. According to VALS research, individuals make purchasing decisions based partly on their personalities.

Additional sources for psychographic information are found at the end of the book. *The Lifestyle Market Analyst* is one publication that correlates demographic characteristics with consumer buying behavior and can be very helpful in your research.

Perhaps your best source of information for specific data (both demographic and psychographic) for your local consumer is through observation, interviews, and random surveys. Get to know the local clientele through community organizations (country clubs, churches, charities, for example). Talk to the local professionals in your ***Chamber of Commerce, Better Business***

Bureau, Retail Merchants Association, and others. Best of all, make a friend of a local store owner who attracts a customer similar to the one that you will want to entice. This is usually quite doable for students, as most local merchants will be pleased to talk about themselves and their businesses and will want to help students seeking information for educational uses. You will need to be careful, however, about the way you ask questions that require the divulging of possibly confidential information. Instead of asking "What is your annual sales revenue?" you might ask "What do you feel is a reasonable sales revenue plan for a store such as mine [give the owner some details]?" or "How many customers would you expect to buy something on a typical day in a store such as mine?"

Those of you who are using this text to prepare a plan for an actual store opening need to be a little more cautious about what you ask and whom you ask to give you information. You will want to avoid polling the owners who are your direct competitors and opt instead for a store owner in another similar town or on the other side of your town or city (if you will not be competing directly for their sales dollars). Trade conventions such as MAGIC in Las Vegas are good places to meet store owners in other parts of the country who will probably be willing to share information with you.

Informal polls of potential customers can also be helpful, but they must not be used as the final standard. Only a formal survey under controlled conditions can give you truly accurate data about potential customers.

Business Structure

It is important that you choose a business organizational structure that works for you and any partners you may have. For a classroom simulation with only a semester for completion, you will likely want to choose a partner from among your classmates. If you are an entrepreneur, you may choose to go it alone or to team up with one or more partners. In any event, you should explore the organizational structures and their advantages and disadvantages:

- A sole proprietorship is the most simple and common form of business ownership and involves one person operating as the single owner. That person will likely have others who will assist and work in the business, but he or she will be the only one responsible for final results and will be the ultimate decision maker. This structure is easy to set up and to operate, but it can be risky for the owner, as all debts have unlimited liability and all ***profits*** are taxed as ***income*** to the owner personally.
- A general partnership is a business structure that is formed by two or more partners who are equally responsible for the management, risk, and debts of the business. This is also a fairly easy structure to establish and requires only that the partners register the business name with the county/city clerk's office in their locale. It is recommended, however, that partners in this type of business have a lawyer write a formal partnership agreement to outline responsibilities, risk management, and other important concerns. All profits for the business are taxed as income to the partners.
- A limited partnership is similar to the general partnership, except that there are two types of partners in this enterprise: a general partner has a higher percentage of ownership and more control over the decisions about the business, while limited partners have less of each. General partners have no limits on risk or reward (profits), while limited partners receive only a share of profits based on percentage of ownership. Liability for these partners is similarly prorated.
- A "C" corporation is formed by partners or individuals who have received a charter recognizing the business as a separate entity from the partner(s) who own it. It is a complex form of ownership in that it involves shareholders, directors, and officers and is more formalized than the structures explained above. The corporation can own assets, borrow money, and conduct business without direct liability on the part of the owners. This format is subject to

more government regulation, and corporate earnings are subject to "double taxation" when both the corporation and the stockholder dividends are subject to taxation. Advantages are limited liabilities to personal assets of partners (but not total protection from lawsuits) and some tax benefits. Disadvantages are that these structures are both expensive and difficult to set up and maintain.

- The limited liability corporation, or LLC, is a newer, quite popular form of business ownership. It allows the formation of an enterprise that legally operates separately from the personal assets of the owners. It also provides the benefit of limited liability for the owners personally, as these owners risk only their original investments. Advantages are that it provides some of the benefits of a corporation but is easier to set up and maintain.
- A subchapter "S" corporation is related to the LLC and can be set up as a corporation for either a sole proprietor or a partnership. The advantages are that it provides for a more simple tax structure than a C corporation (which is more complicated and more expensive to establish and maintain) and provides liability protection to the owner. This form of ownership allows income to pass through to the owner, where profits will be taxed at the individual rather than corporate rate; and it has fewer restrictions than an LLC.

(*Note:* Information for this list is from Richmond, Virginia, U.S. Small Business Administration, *Small Business Resource Guide*.)

According to Leroy Keller, Lender Relations Specialist for the Richmond, Virginia, Small Business Administration, the most common type of structure selected by small business owners who own no real estate for the business is the subchapter S corporation.

There are other forms of ownership with which your lawyer can help, but these are the most common. For the simulation, you will need to choose one of these (or another that you have researched) and state this when you write your Business Description (Part IV of the business plan outline).

Simulation

1. Begin by addressing Part V of the business plan, which is the Product and Service Plan. Write a brief discussion of the product line, list your merchandise categories, and describe your store niche. Use your research on the industry and product line to help you with this section. Consult the resources in the Introduction and at the end of this book to help with your research. Remember that you should not do the business plan in a question-and-answer format, and be sure to address all of the following issues/questions:
   a. What are you selling: product categories, key retail price points, key vendors?
   b. Why will your target customers come to you?
   c. What is different about your product or service? Identify your niche.
2. Using the business plan outline given in the Introduction, complete the Customer section of Part VII Marketing Plan: demographics, psychographics, and other pertinent information about your target customers. Also address the potential for your product category in your particular market. Be sure to address the following issues:
   a. Identify your potential customers by demographics utilizing information from several different sources (see those listed in this step and in the Resources; include ***census*** data).
   b. Define the target customer that you intend to pursue. Discuss the customer's presence in your market area (How many? Where do they live? and so on).
   c. Is the market (both nationally and locally) for your products growing, steady, or declining?
   d. What is the projected growth potential for your product in this local market? What is the potential growth for your store?
3. Remember that this is just the beginning. You will have the opportunity to revise or add to this information when you finalize the product, marketing, and

business description in later steps. After you complete the first step of this list, file it in a notebook or special file titled Part V: Product and Service Plan. The work you complete for steps 2 a and b of this simulation should be stored in a file titled Part VII: Marketing Plan. Finally, your work for sections 2 c and d should be put in a file marked Part IV: Business Description. You will use this information later when you complete these three sections. Right now it is important to establish who your customer will be and what your store will do to draw business away from the competition.

STEP 2
Finding the Right Location

"Location, Location, Location . . .
Where have I heard that before?"

PROJECT OBJECTIVES

This step in the simulation will guide you through:

- Working with a commercial real estate agent
- Analyzing the benefits to store location and environment
- Evaluating the key information and expenses in choosing the right store location

Anyone who has studied business or thought about starting their own business has no doubt run into the clichéd phrase about the three most important things to be addressed in determining where to physically place the business. Those three things are, of course, *location, location, location*. This tried-and-true mantra has probably appeared billions of times as the answer to a question on business professors' exams and uttered even more often from the lips of commercial real estate agents. Naturally, when making the location decision, the business owner has to factor in a myriad of considerations, and it is much easier to say that location is important than actually going out and finding the right one.

Working with a Commercial Real Estate Agent

It has been our experience that when trying to make a difficult decision, it is best to gather as many facts as you can about the issue and then find an expert or experts who can help you make the ultimate decision. There are no doubt hundreds of commercial real estate agents in the community where you live, and it would be wise to take advantage of them. They know much more about the available business real estate than you probably ever will and have the necessary resources to help you evaluate it. Most important, if you are on the buying or renting end of the deal, then their services are virtually free. Remember, the seller/landlord is the person who pays the real estate agent when the deal is done.

Finding the right agent is not as daunting as it may seem. There is lots of help for you here, too. The best and probably most reliable way to find a good agent is to ask around and get recommendations from your circle of family, friends, or business acquaintances. These people know you and they no doubt would know the type of agent who would fit your personality. After all is said and done, you want a person with whom you are compatible and one with whom you can get along because you are going to spend a lot of time together.

Real estate agents come in all shapes and sizes. That is, they all have different levels of expertise, and you will want one that specializes in your type of business. You also want one who has both a good track record and a sterling reputation. Check with your local better business bureau or chamber of commerce to get lists of licensed agents. Of course, the best place to start is the Internet. When you type the phrase "commercial real estate" into a popular Internet search engine, over 4.6 million sites are listed. The Resources at the end of this book lists the most popular sites.

Buy vs. Rent

Do you have oodles of money lying around? Do you have a place that someone is going to give to you for free? Do you have a rich uncle (aunt, mother, father, brother, sister, grandparent) who will fork over the dough you will need? If you are like nearly everyone, then the answer to all of the above questions is a resounding no!

With the limited budget that most small businesses have when they start, the decision on how to pay for the business location is a very big one. Tying up a huge portion of your ***liquid assets*** in purchasing a building is probably not going to be a very good idea. Not only will the ***cash flow*** be severely affected by what will no doubt be a significant down payment, but the owner will also be responsible for the maintenance, taxes, insurance, and upkeep of the building. Remember also the hard fact of life is that most businesses do not make it beyond the first year. Having to dispose of real estate would place an even greater burden on an already stressful situation.

Naturally, one size does not fit all, but unless you can negotiate an extremely good deal or have easy access to a location, in our judgment renting is probably the best option for the new business. Renting gives the business owner the opportunity to "test drive" the location. It allows for tweaking and fine-tuning during the birth pains that all businesses go through. Should the location not work, it is no doubt easier and quicker to negotiate getting out of a ***lease*** than it would be to try to sell a purchased location.

Types of Leases

It goes without saying that a real estate lease is one of the most important documents that the business owner will ever negotiate. Many factors come into play when determining which lease is best. Therefore, the owner is wise to find and heed the advice of his or her real estate agent or attorney. A section later in the book discusses when and whether the business owner should seek legal advice, but it is definitely a worthwhile expense to get the advice of a competent real estate attorney when negotiating a lease. Real estate law is too complicated for the novice to waste time trying to read and understand all of its nuances.

According to http://allBusiness.com, the following are the main types of commercial real estate leases. From a broad perspective, there are a few types of leases commonly found. Within these categories, leases may vary considerably.

- *Gross lease.* The tenant pays a set amount of rent and the landlord is responsible for payment of taxes, insurance, and other costs associated with owning the property.
- *Net lease.* The tenant pays the rent plus a portion of the maintenance fees, insurance premiums, and other operating expenses.
- *Triple-net lease.* Typically, for a freestanding facility, this type of lease has the tenant paying for all fees and operating expenses associated with the space.
- *Shopping center leases.* The tenant pays a base rate in conjunction with the square footage of the retail facility. Typically, the tenant also pays some common charges and frequently a certain percentage of the gross sales. The tenant may also be assessed part of the property taxes. A shopping mall lease often includes terms about signage, hours of operations, common areas, and deliveries. The landlord may also have the right to relocate the tenant.
- *Land or ground lease.* The tenant leases the grounds and builds on the property. Typically, with a land or ground lease, all improvements on the property, including any building or buildings, revert back to the landowner at the end of the lease period.

Matching Product Line to Location and Type of Store

One of the most important considerations in selecting a site for your new store is whether or not the store location is compatible with your product line and ***price points***. You should first determine whether your store will be a "destination" store; in other words, will customers make a special trip to purchase your products or must you depend largely upon customers who are already shopping in your area? The answer to this question will largely determine the type of location that you should choose. If your store is truly a destination store, you will have much more flexibility in location; perhaps you could even locate your store in the basement of your home (if zoning regulations allow). Unfortunately, this is not often the case with fashion-related stores; they need other stores around them that will attract the same target customer who will likely be on a shopping trip, perhaps with friends or family, or just out for a look at what is new.

As you can see from Figure 2.1, Rumors is not your typical women's boutique. The two young women who opened the store are recent graduates of Virginia Commonwealth University, and they wanted to open a store that would appeal to trendy college women. (See CD-ROM for an interview with the owners.) What better location than a store right in the heart of the VCU Academic Campus? Although there are no other clothing boutiques near them, they are surrounded by places that college students frequent: a café, a coffeehouse, and other places to get a quick lunch.

Figure 2.1 *The Rumors store interior has a unique image and merchandise.*

Remember that your location is a big part of your image. Most of you will be opening a boutique selling some kind of fashion goods and, as a small store, your price points will probably be higher than the average; perhaps you will be selling luxury goods or high-fashion products targeting a higher-income shopper. In any event, the most important consideration is to match your location to your product line and target customer base. Decide where they are most likely to shop and what kind of store and location would best suit their tastes. Although we will stress throughout this text the importance of controlling your expenses in a new store start-up, it is also crucial to use good common sense in such an important decision as where to locate your store. As the adage goes, "Don't be penny-wise and dollar-foolish"; in other words, don't pick a store site just because you feel that it is cheap. If you are aiming for shoppers who want better or luxury goods, don't expect them to come to a section of town that caters to lower-income or bargain shoppers. Locate your store in an area that they already frequent.

In Step 7 we will thoroughly explore your competitors, including their strengths and weaknesses; but before we make that analysis, you need to be aware of what kind of stores will be your neighbors. Ideally, the stores around yours will help to draw your target customer and make sure that he or she is in the neighborhood on a regular basis. Compatibility is key; do not open a fur store with a natural, environmentally friendly boutique on one side and a blue-collar-oriented clothing store on the other. You will only confuse target customers. You also do not want to open a store that is almost a duplicate of the one located right next to yours (see Introduction about creating your own niche). If you intend to open a shoe store in a strip center and there is already one, make sure that your shoe store offers different products or has a different niche. It is always advisable to make friends, not enemies, of the surrounding stores; hopefully, you will work together on promotions, advertising, and other joint efforts to draw consumers. In many cases, landlords or realtors have regulations preventing duplicate stores from opening in the same centers.

When searching for the right location, don't grab the first store that you see. Look at several sites for comparison before settling on the best choice for your product. Be wise about costs. Keeping in mind what we just discussed about marrying location to target customer, you also don't want to obligate yourself to such a high rent that you can't make a profit. A good guideline for rental cost is that totally it should represent about 7 to 10 percent of your annual sales plan (Base rent + Common area expenses + Any surcharge for sales performance). Even before you finalize your sales plan, you should be able to reason that a store in a high-rent district requires a product line that will produce high sales volume. Also, remember that a good location might result in lower expenditures on marketing, whereas a poor location will likely require a larger advertising budget to pull in the customer.

When shopping for location, do some people watching. Observe the kinds of potential customers who are walking around, shopping, and working in the area. Analyze your observations in light of the customer that you need for your store. Evaluate the shopping habits of potential customers: women often like to shop on their lunch breaks or right after work and enjoy stores that are located next to the workplace; men more often shop where and when they can combine other activities such as seeing a movie, eating in a restaurant, or shopping in another store for items such as technology products or sporting goods.

You would also consider the following factors when selecting your site:

- Is there space enough for proper display and housing of your product categories—conversely, is there too much space? Maximizing sales per square foot (discussed further in Step 3) is crucial to making a profit. In other words, space costs money and you don't want a lot of empty space in your store that is not productively used.
- What is the profile of customers living near the store? Don't open a very trendy children's boutique in an area that caters primarily to senior citizens.
- Is there foot traffic in the area?

- Will you be required to pay common-area fees for maintenance and upkeep of the area? These are usually charged in a per-square-footage fee to cover such things as repair of common store areas such as roofs and sidewalks and upkeep such as mowing, cleaning, and snow removal.

Benefits to Store Location and Environment

The importance of the physical location of your business is dependent entirely on the type of business. Businesses that depend on telephone or Internet sales can be located almost anywhere. Similarly, a consulting type of business may not even have a physical location because the business is almost always conducted where the client is based. However, the location of a retail business is crucial because the success of the business is based on how many customers come through the door.

Figure 2.2 *The exterior of the Heidi Story store.*

For example, as shown in Figure 2.2, the fashionable women's clothing boutique Heidi Story is not only located in the trendy Carytown, Richmond, Virginia, shopping district surrounded by popular eateries and complementary stores, it also offers a very inviting entrance to entice the customer into the store. The large awning is a soft, feminine pink, and the windows and large door offer a welcoming view of the store's merchandise. (See CD-ROM for an interview with the owner.)

In addition to an appropriate physical location of the retail business, it is vitally important that the business also provide an environment that the customer will find attractive and inviting. Areas that the potential businessperson must consider include the following:

- The storefront of the building must be welcoming and accessible.
- The layout of the store must be easy to step in and customer friendly.
- The colors on the walls must be appealing and match the personality of the store.
- The merchandise must be easy to access.
- Parking must be easy, accessible, and ample.
- Salespeople must be friendly and knowledgeable.
- Signage must be prominent and easy to understand.

Get the idea? Everything about the store must be customer friendly. If the customer is not happy, then rest assured that you as the business owner will definitely not be happy. Every decision that is made must be made with the customer in mind.

Naturally all of the above has to be factored into what the business owner can afford and what the real estate market has available in its business inventory. This balancing act of "Where can I be the most successful?" and "What can I afford?" is crucial. The business owner cannot—make that *must not*—start a business in a locale that will not support the expenses of the business.

A good business location checklist can be found at http://smallbusiness.

findlaw.com/business-operations/commercial-real-estate/be4_4_1.html. The small business owner can also consult his or her real estate agent for advice, as well as the local chamber of commerce, small business development center, or trade association.

Key Information/Expenses to Evaluate in Choosing Your Store and Location

After you find your store location and meet with the realtor, and as you are signing the lease, you will want to make sure that you have all the pertinent information needed to evaluate your potential and begin the process of financial planning. Much of this information will be on the lease; however, every lease is different and not all contain all the information you will want to know. We recommend that you fill out a form that lists the important facts about your store and your lease. This form will serve as a quick reference for anyone who is reading your plan and does not want to read the entire lease.

Table 2.1 is a completed form with information from our simulated store in Carytown. Our store was previously an antique store that occupied both floors of a building that was built as a residence in 1922 and converted to a retail store in 1956. Carytown has more than 300 stores primarily made up of buildings that were at one time houses, banks, churches, or similar structures. Since 1956, several structural changes to our building have resulted in a large space of over 7,200 square feet, two floors, and many small, disjointed rooms.

Knowing that we would need to provide in our business plan a map pinpointing our location, our realtor provided us with the map shown in Figure 2.3 of the Carytown area. This map will help the reader (a financial backer, for example) get a better picture of the area surrounding our intended store.

Our first visit to the site could have been discouraging, as the previous tenant was in the midst of a closeout sale and the store was filled with assorted

TABLE 2.1 STORE LOCATION FACT LIST (completed)	
Store location: 3445 West Cary Street	
Cross streets	Cary Street fronts the property, and Nansemond Street is the nearest cross street
Dimensions of entire space: Total square footage	48-foot frontage x 62.5-foot depth = 3,000 square feet (after renovation)
Dimensions of separate spaces: Square footage of bathrooms, square footage of store rooms	Total square footage of non-selling space (office, storage/stockroom, bathroom plus planned area behind cash/wrap desk) is 800 square feet, leaving 2,200 square feet of planned selling space.
Shape of store	Renovated first floor will be a triangular space with a small indent at right back side of store.
Location of entrances/exits	Two entrances to the building will be on the front of the store: On the far right side will be the entrance to our shop, and on the far left will be the entrance to the stairs and upper-floor tenant.
Locations of windows	Double large windows at right front of store next to door. Large bay window at left front of store.
Monthly rent, Rental/square foot	$17 per square foot. $51,000 annual, $4,250 per month.
Terms of lease	Three-year term, triple net lease.
Common area fees: Maintenance, marketing, etc.	No local merchant's group fees, no maintenance.
Security deposit	Two months' rent.
Realtor fees	Paid by owner of building
Expenses incurred by lessee: Utilities, etc.	Est. fuel oil costs for renovated portion: $750/year. Est. electricity: $4,000/year. Any future interior renovations (paint, carpet, etc.) to be paid by lessee. Property insurance to be paid by lessee.
Allowable renovations	Must be approved by landlord. No renovations planned in addition to those being made by owner; lessee will paint, install new carpet, and add dressing rooms.
Landlord obligation/commitment for renovations	Has agreed to renovations as specified in floor plan drawings. Future: Landlord pays exterior upkeep as well as upkeep to interior plumbing and HVAC unit.
Miscellaneous	Current bathroom is small; plans in future to expand or add to bathroom(s). Plan some built-in fixtures.

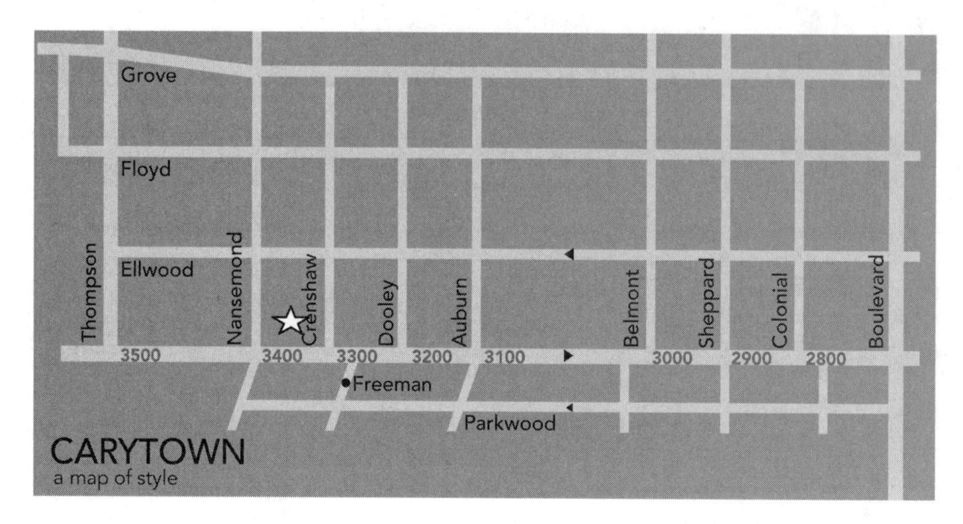

Figure 2.3 *Map of the Carytown shopping area.*

merchandise and boxes and was in general disarray. If you would like to take a look at the interior of the simulated store the way that it looked on our first visit, check out the photos on the CD-ROM. This situation is not unusual for a potential new store owner, and one needs vision in order to imagine what can be done with the structure. Also, as is often the case when you are taking over from a previous tenant, we could see that the space had many advantages, but also disadvantages. Advantages were primarily that it is in a great location to attract our target customer, the exterior of the building was well kept (see Figure 2.4), and while the interior needed some renovation, it could easily be turned into a charming shop with interesting mini spaces for display and housing of our merchandise. Disadvantages were that the total space was too large for practical use for a small clothing boutique, and it needed some cosmetic as well as structural work.

Since our vision for our simulated store is a ladies' retail clothing boutique, we knew that the entire space was too large and too expensive for us to rent, and many of the rooms were too small to be viable for our plan. We approached

Figure 2.4 *The exterior of the Martha's Mixture store.*

Thalhimer Commercial Realty, the firm that is representing the rental of this property and one of the largest firms in the Richmond area. We asked them to propose some changes to the owner of the building. We only needed 3,000 square feet of space and only one floor. We wanted to have some walls taken down to maximize our ability to properly merchandise our selling area. Keeping in mind that *this is only a simulation* (the store in question is not actually being renovated for us), note the floor plan sketches in Figures 2.5 and 2.6 (also on CD-ROM) and how they show the changes that the owner has agreed to make for us to have a usable space. The advantages to the owner are that he will be able to rent two additional spaces and will also have a 3,000-square-foot space that is much more usable for a retail boutique.

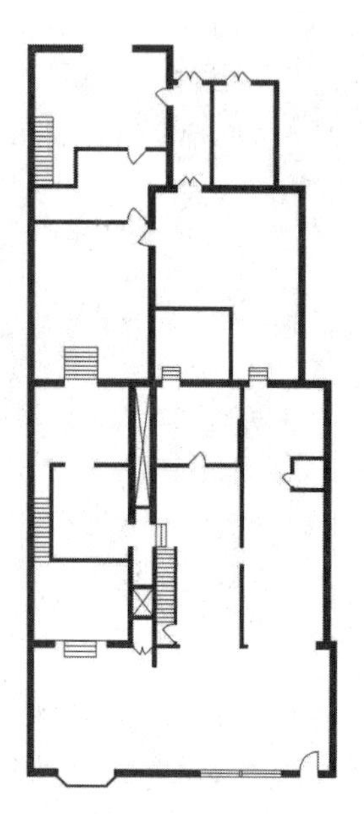

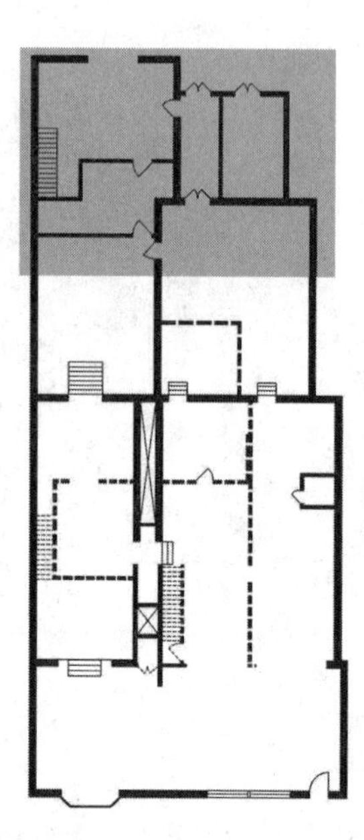

Figure 2.5 *Original floor plan sketch.* **Figure 2.6** *Demolition floor plan sketch.*

Finding Help to Plan the Renovation

Figure 2.5 shows the first-floor layout as we saw it on the day of our first visit. Figure 2.6 is the demolition plan that we developed with our interior design consultant, Jennifer Hamilton. We found Jennifer through the Virginia Commonwealth University Department of Interior Design. She is a recent graduate of the master's program and an assistant professor in the department. Jennifer suggested that we ask the owner to agree to move the staircase on the left side of the store to the front of the store so that the upstairs could be leased to another occupant. We also asked them to close off the staircase in the center of the store, as neither we nor the upstairs tenants would need it. It was noted by our consultant that the elevator in the middle of the store would need to remain for purposes of safety, people with disabilities, and so on, but it would not be used during our tenancy. We also asked the owner to wall off the back of the store for another use (separate tenant, his storage, etc.) and put in an

ATM with access from the street. You will note from the plan, however, that we had to put in an exit door through the back of the store that would allow access in case of emergencies. All of these changes would leave us with approximately 3,000 square feet for our store.

Further work with our consultant Jennifer gave us some additional great ideas for making our store customer friendly and very usable for displaying and selling our product. She created a layout with three dressing rooms in the rear of the store near the area that would be used as our stockroom (see Step 9).

At this point, two items should be noted. First, not every owner will agree to the number of changes that we have simulated for our store. In some cases, he or she might not want to change the structure and layout of the store. In other cases, the owner might say OK but might want the lessees (us) to assume the costs. For this simulation, we are assuming that we were able to convince the owner that these changes would be financially beneficial. Many first-time store owners have to accept the property as is and work around structural challenges. Of course, any changes that affect the structure of the store itself (supporting walls, etc.) cannot be made. Second, while we will continually stress throughout the book the importance of controlling expenses in order to make a profit, it should be noted that what you do structurally at this point will stay with you for a long time. Don't go crazy with spending money on renovations, but make sure that (like your merchandise) the look of the interior of your store will appeal to your target customer and marry well with your merchandise assortment. While you might be able to plan your interior alone, it is worth considering the help of an interior design consultant like Jennifer. We decided to spend the extra money for Jennifer's help (approximately $1,000 in our case) so that we would have an interior to be proud of and so that we could better convince the building owner of the wisdom of our suggestions. We had Jennifer do all the renovation plans and design an interior layout of both built-in and purchased fixtures (shown and described in Step 9). If you need more changes, you might consider using the services of an interior design or remodeling com-

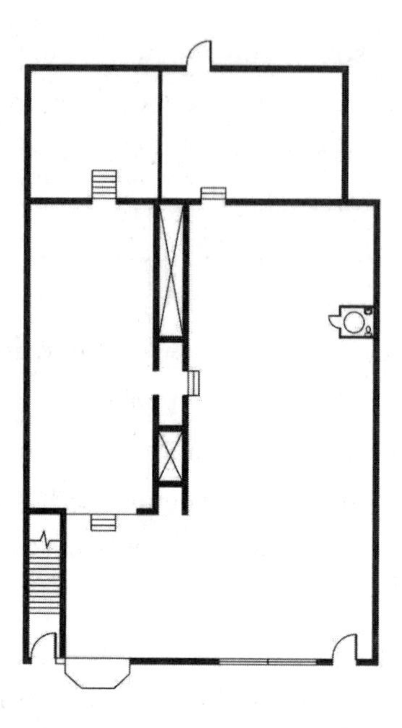

Figure 2.7 *Reconstructed floor plan sketch.*

pany; fewer changes might mean that you will want to consult a fixture company for suggestions and help or do your own research.

Figure 2.7 shows how our store will look after all renovations have been made. At this point, we are ready to plan our merchandise assortments and then determine our fixture layout. We will plan our product assortment in Step 6 and our floor plan and fixtures in Step 9.

Rental costs in the Carytown area typically run from $15 to $20 per square foot. This is a method by which realtors and landlords determine annual rental costs. They use standards that exist for a particular part of the city for the dollars that will be charged per *total* square footage of the property being rented. That standard can vary some from space to space within the assigned area, depending upon use of the space, size of the space, and other factors. In the case of our simulated store, we are being assessed a number that is in the middle of the standard.

For a 3,000-square-foot store, we might have been able to negotiate the lower price of $15 per square foot, but the landlord is spending quite a sum of money on our suggested renovations, so we must pay a little more because of this. Using the $17-per-square-foot price and multiplying it by the total square footage of our space, we arrive at an annual rental of $51,000 with a monthly rental of $4,250.

Most of the categories listed in Table 2.1 are self-explanatory; you can use the Miscellaneous box to include any important information for your particular location.

Simulation

1. Find an empty location somewhere in your immediate area. For students, this should be a locale near your college or university. Entrepreneurs: You probably already have a plan for your desired city; now use the principles discussed earlier in this step to find the best site for your business.
2. Fill out Table 2.2, which also appears in the Blank Financial Documents folder on the CD-ROM, with the pertinent information for your location.
3. Fill out a lease agreement for your store location. Entrepreneurs: You can get the correct form from your realtor, and he or she will fill in the information and take charge of necessary signatures—yours and that of the landlord. Students: You will probably be able to get a blank lease from the commercial agent responsible for the rental of your store; if not, you may use the blank lease, which is located in the Sample Blank Lease folder on the CD-ROM. You may also download a free commercial lease form from any one of a number of Web sites. Students, make up a name (John Q. Public or similar) and simulate the signature of the potential landlord for the purposes of this project. Both you and any partners you have will need to sign and date the lease also.
4. Find a map of the area surrounding your store and be prepared to include that with Part VI Location of the business plan.

TABLE 2.2 STORE LOCATION FACT LIST (blank)	
Store location:	
Cross streets	
Dimensions of entire space: Total square footage	
Dimensions of separate spaces: Square footage of bathrooms, square footage of store rooms	
Shape of store	
Location of entrances/exits	
Locations of windows	
Monthly rent, Rental/square foot	
Terms of lease	
Common area fees: Maintenance, marketing, etc.	
Security deposit	
Realtor fees	
Expenses incurred by lessee: Utilities, etc.	
Allowable renovations	
Landlord obligation/commitment for renovations	
Miscellaneous	

5. Write a description of your location that addresses the following issues:
    a. How will this location contribute to the success of your business?
    b. What benefits to the location will appeal to your target customer? For example, is it easily accessible? Does it have sufficient parking? Is it a safe area with good street lighting? Is public transportation available?
    c. Address any problems associated with the location and what you plan to do about these issues.
    d. This discussion will become part of Part VI of your business plan, and you will be asked later in the simulation to combine this information with some additional data and roll it all into Part VI.
6. Create a floor plan that shows the important elements in your store: walls, doors, display windows, bathrooms, office space (existing or planned), stockroom (existing or planned), and other structural parts that may impact the visual merchandising of your store (indicate, for example, if there is a support beam or post somewhere in the interior of the store). You may create this floor plan using Adobe Illustrator or CAD, or any other program that will allow you to make a floor plan. The folder Illustrator Instructions on the CD-ROM provides instructions for creating a floor plan using Adobe Illustrator. When you have finished this plan, it will be ready for the work that you will do in Step 9 when you place your fixtures and other visual merchandising elements.

STEP 3
Funding the Business/Sales Planning

"You want Uncle Sam . . . No, Really!"

PROJECT OBJECTIVES

This step in the simulation will guide you through:

- Types of ***investors*** and where to find them
- Pros and cons of investor types
- Working with lending institutions
- Types of loans and the role of government
- Formulating realistic sales goals and fixed expenses
- The role of accountants and lawyers

Ways to Fund Your Business

It has been said that money is the root of all evil. Well, while that axiom has been debated for thousands of years, when it comes to owning a business, money is the root all right, but it is certainly not evil. Money is the lifeblood of our capitalistic society, and it is most definitely the driving force behind successful businesses. For a business to succeed, it must make a profit. To make a profit, it must take in more money than it doles out. Simple, right?

Businesses fail for many reasons, but one reason that is right at the top of the list is undercapitalization. You may have the best business idea in the world. In fact, you may have indeed invented that better mousetrap that everyone seems to be wanting. However, if you cannot fund the business to build the mousetrap, market it, and sell it, you may as well have just spent your time playing an accordion on the street corner hoping for handouts from passersby.

Hundreds if not thousands of resources are available to the business owner in search of ***capital*** (Figure 3.1). It is not only important for the business owner to know where to go for help in obtaining capital but also to know how much capital is needed. In addition, before approaching a potential investor, the business owner will need to have a comprehensive business plan available for the investor's review.

Business investors come in all shapes and sizes. They are available to the smallest mom-and-pop store and to the biggest multinational corporation. Their purpose in life is to take their investment money and turn it into more money. They are very picky about to whom they entrust their money, and they are very unforgiving when their money is not used wisely or in ways other than was promised to them. It, therefore, behooves the business owner to be extremely forthright and honest with his or her investors.

Types of Investors and Advantages/Disadvantages of Each

For this book, we have broken down the types of investors into three basic categories:

- Personal
- Lending institutions
- Government

We have not attempted to list every known kind of investor but are highlighting the most common types. The Additional Resources on the CD-ROM will help the reader find more detailed information on investors.

Personal Investors

Personal investors are those usually known by the business owner. They generally are family, friends, and acquaintances. These types of investors are usually the most reliable and easy to deal with because the owner knows them so well. They are more forgiving when payments are late and in fact may even be in a position to wait for payback until after the business is up and running or to even forgive the debt if the business fails. In addition, personal investors usually require no collateral for the money they lend. Also included in this category of investors is the owner him- or herself.

There are also difficulties presented by the personal investor. The number of resources is likely to be limited and, accordingly, the amount of money available will also be limited. In most cases, the initial investment will probably be a onetime deal and once tapped cannot be used again. The business owner must also assess the risk of the impact of the business's failure on the investor. We have all heard that loaning money to family or friends can be a dicey proposition. If the person asking for the money cannot afford the loss of a good relationship with the family member or friend, then it is best to not use them as an investor in the first place. It also goes without saying to make sure that all agreements involving investments from personal resources are put in writing.

Lending Institutions

By far the most used and most widely available sources of funding for small business start-ups are the many lending institutions that one can find on almost every street corner. They are plentiful, widely available, have plenty of money to lend, can cater to the smallest business or to the largest, and are very adaptable to changing circumstances, and they can be used over and over and over.

In the case of small-town banks, it is likely that they know the business owner and can often provide the same kind of advantages that a personal investor can provide. Large banks often have local branches that are designed to serve the small business owner in the same way as the local bank.

Credit unions are another source of capital for the business owner. If the owner is a member, then he or she already has a relationship with the organization. Many times their interest rates are better than the traditional bank, and

Figure 3.1 *The entrepreneur can obtain funds for a new business from a number of sources.*

because the members run credit unions, they are more sympathetic to the business owner's (member's) needs.

Another source of investment money for the business owner is credit cards. These funds are readily available, and the owner usually has an already established credit line. However, their interest rates are usually very high and the amount available to the borrower is limited.

Investment money from lending institutions can be difficult to obtain. They all require a large amount of paperwork in the form of business plans, financial history, and personal information. Banks are also very conservative in nature, and their loss rate is extremely low, so they are very careful with the money they lend. A standing joke among business owners about banks is that they are always willing to loan you money when you don't need it and are never willing to loan you money when you do.

Government Lending

The government's role in lending to businesses is somewhat unusual. In almost all cases, the money is not loaned directly from the government. Most financial aid to business from the government is in the form of a loan guarantee. The government will allow regular lending institutions to lend the money to the business, and the government may then guarantee that loan. In other words, if the borrower fails to pay the loan back, then the government will reimburse the money to the lending institution and the borrower will owe the money to the government. These kinds of loans are in place primarily to assist start-up businesses, which historically have a more difficult time obtaining the loans because of their risky nature. The business owner needs to be aware that these loans are just like regular loans and may have to be collateralized and must be paid back. The SBA is the primary federal government agency that provides these kinds of loans.

Working with the Lending Institution

One constant theme that runs throughout this book is that planning is the paramount element of a successful business. Nowhere does planning play a more important role than when approaching a lending institution for a loan. Not only do lending institutions expect to have the money they lend repaid to them but they also expect it to be repaid on the schedule that they set up.

Before the business owner approaches a lender, he or she must have prepared a business plan that is detailed, reasonable, accurate, and understandable. The lender should be able to look at the plan and know exactly not only how much money the business will need but also how it is going to be spent and how it is going to be repaid. The business plan is the blueprint of the business, and it is the most important document the lender reviews when determining whether or not to approve a loan.

Most lenders also require that business owners show their commitment to the business by putting their own money into the business. How much depends on the lender, but know that if the owner does not have enough faith in his or her business to invest in it, then no one else will either. The more the owner can invest, the less debt the business will have, and the less debt the business has, the more attractive it is to lenders.

Another element that may make it easier to obtain a loan is to put up collateral to back the loan. Collateral can be anything from a car to the owner's home, to bank deposits, to cosigners on the loan—anything the lender considers valuable. Most lenders, however, do not want to be in the business of selling ***collateral*** to pay off a bad loan. If collateral is involved in a loan, it usually must be something that the lender can turn into money with little effort.

There are also intangibles that go along with being successful at obtaining financing from a lending institution. Remember, these guys are generally very conservative and, rightfully so, are zealots at protecting the assets of their institutions. When business owners approach them for money, they must sell themselves as well as their business.

Owners must present themselves as knowing what they are about. They must show themselves to be of good character whom the lender can trust and depend on to be valuable and trustworthy clients for the institution. It is also valuable to already have a relationship with the lending institution, and the borrower's personal ***credit rating*** needs to be as high as possible.

Types of Loans

Several different types of loans are available for small businesses. Knowing the correct type of loan to ask for is essential in getting the bank to approve your loan. There are two basic categories of loans: secured loans and unsecured loans.

Secured Loans

These loans are secured by collateral. The lender will take a security interest in your property. If you do not pay the loan back, the lender has the right to seize your collateral. Most lenders require collateral to secure a small business loan, and they may take a security interest in your business and/or personal assets. Lenders will not lend you more than 100 percent of the value of your collateral and will usually only lend you 60 percent to 80 percent of its value.

Unsecured Loans

As the name implies, these loans are not secured by any collateral. Credit cards, while not technically loans, are the most common example of unsecured debt. The lender is loaning you money based on your reputation and credit worthiness. It is very, very rare to find a lender willing to give you an unsecured loan for a new business. The lender is not willing to take the risk because you have no track record for them to work from.

Typical Bank Loans for Small Businesses

Following are the typical loans banks will provide to small businesses:

- Unsecured credit lines. Banks and lenders will often provide lines of credit to small businesses. These are generally set as a maximum amount of money you can borrow. The credit line amount and interest rates vary greatly from lender to lender and from business to business. In general, the bank will look for the following three factors:

1. Does your business have its checking account with the bank?
2. How long has your business been in existence?
3. What are your personal credit score and your business's credit history?

Most lenders will also provide you with a business credit card in addition to a line of credit.

- *Short-term loans.* Short-term commercial loans may provide a good source of ***working capital*** for your business. The length of these loans is usually no more than three years, and the loan will require fixed payments of principal and interest. Short-term loans will need to be secured by adequate collateral and will usually have a fixed interest rate because of the short length of the loan.
- *Long-term loans.* Long-term loans are almost exclusively used for equipment and other ***asset*** purchases. Lenders will not lend your business money for longer than three years unless the loan is for a specific asset purchase or for the refinancing of an existing asset. The assets being acquired secure these loans and will generally have various loan covenants such as interest rate changes and prepayment penalties associated with them.

The Role of Government

When it comes to determining whether or not a business will succeed or fail, the big elephant (or donkey) in the room that cannot be ignored is the role of government. We are talking about government at all levels here—federal, state, and local. All have their pluses and minuses, but none of them can be ignored. In fact, whole books have been written on the subject of government and business, but our purpose is to show how government can be a valuable partner in your business.

Local Government

By far most contact with government by the business owner will be at the local level. The city or county government will issue business licenses, conduct assessments for tax purposes, and issue zoning regulations. Local governments may also provide resources to assist businesses in the form of economic development authorities and small business development centers or provide tax incentives for businesses to locate in their area.

State Government

In most cases state government will mirror the local government's role except at a more powerful level. While the state will also have its economic development authorities and small business development centers, they also may have loan programs to assist businesses.

State government also controls the regulatory agencies to which business must respond. States make the rules about how to incorporate as well as setting licensing requirements. They set the standards by which businesses are judged and regulated. The list of boards and authorities set up by states to regulate business is almost endless—from trash dumping to funeral homes, to construction standards, to real estate, to nursing homes, and on and on.

Federal Government

It is almost impossible to define all of the elements where the federal government touches business. Every aspect of business has some agency in the federal government that either regulates it, taxes it, provides assistance to it, or sets standards for it. Since this text is about helping the small business owner, it will be limited to the areas of the federal government that most affect it. However, it behooves the business owner to make sure that he or she is well versed in how the federal government addresses his or her type of business. The Internet makes this task much easier than ever.

The agencies at the federal level that have the most impact on business are the Department of Commerce (www.commerce.gov) and the Small Business Administration (www.sba.gov). Commerce defines its mission as "to foster, promote, and develop the foreign and domestic commerce" of the United States. This has evolved, as a result of legislative and administrative additions, to encompass broadly the responsibility to foster, serve, and promote the nation's economic development and technological advancement.

In other words, the Department of Commerce's purpose is to enhance the commerce of the United States. This involves international trade, interstate commerce, economic development, and a multitude of other duties.

As noted earlier, the mission of the SBA is more limited in that its purpose is to "aid, counsel, assist and protect the interests of small business concerns, to preserve free competitive enterprise and to maintain and strengthen the overall economy of our nation."

A quick look at the SBA Web site reveals that the agency is a gold mine of help to small businesses. It provides loan guarantees, business education, business development assistance, assistance with selling goods and services to the federal government, as well as particular development assistance directed to businesses owned by socially and economically disadvantaged individuals.

Developing a Sales Plan

It would be really convenient for the student and the entrepreneur if there was a quick and easy formula for calculating the first year's sales plan, one of the first and most important steps in preparing your financial documents for your business plan. Unfortunately, there is no easy formula, but there are well-tested methods that will yield a workable plan.

Start with Research

As with other parts of the plan development, the most important first step for the new store owner is research. The most reliable yardstick for measuring the accuracy of the sales plan is the results of other stores that are similar to yours. Start your research with gathering information about results of other small stores that are similar to the one that you intend to open. Suggestions for this research are the following:

- Find a small store in your selected location (or one as close as possible to your location) that has a similar target customer base. If you are a student, make friends with the owner of this store and ask for help in gathering data. Potential entrepreneurs might not find help from their direct competitors, but they can ask for information from noncompeting stores, stores in a different but similar town, or stores on the other side of town from your chosen location.
- Spend some time observing customer buying behavior in a similar store. It is best to go back to the same store on different days of the week, different weeks of the month, and different months of the year to track variations in buying patterns. If that is not possible, try to visit the store on at least one busy day and one slow day to test the differences. Your research should involve watching how many customers enter the store, how many actually make a purchase, and how much they typically spend on each visit. Hint: Your new store owner "friend" might give you some of this information as it relates to his or her store.
- Search the Internet and local newspapers and magazines for articles about small stores. Look for information about annual sales results, sales results by product category, and typical customer traffic in small stores. All of this data can be helpful in formulating your first year's sales plan.
- While doing your research, look for available data on sales ***per square foot*** of similar stores. One of the best places to find this data is at www.BizStats.com.

Make Use of Retail Formulas

Sales per square foot is one of the four important keys to planning and controlling your business (the other three are discussed in Step 6 on merchandising).

It is used by stores of all kinds and sizes as a way to measure the success of their sales operation, and as a method of comparing their success against that of their competitors. For the sake of planning first-year sales, one can use this number to build a formula for calculating the plan.

The main purpose of this formula is to help the owner determine whether he or she is getting the most sales for every square foot of space in the store. Understanding that the costs of running a store (rent, utilities, payroll, etc.) represent a big monthly commitment on the part of the owner, he or she would want to be certain that merchandise purchases justify the space that they occupy in the store. There are two different ways to address store square footage: ***Total square footage*** includes everything inside the store; ***selling square footage*** is the total minus non-selling areas of the store, such as stock/storage rooms (except for shoe stores, most of whom keep the majority of their shoes in a stockroom), bathrooms, office space, dressing rooms (unless they have mirrors that aid selling and are then considered part of selling space), and the area behind the cash register. The formula for calculating sales per square foot is as follows:

Net annual sales ÷ Selling area in square feet = Calculating sales per square foot

Develop the Plan

Armed with the results of your research, you are now ready to formulate your first-year sales plan. For most accurate results, take the following steps:

1. Using the information gleaned from your store observations, determine the typical number of customers who would likely make a purchase in your store in a "typical" day (understanding, of course, that there is no such thing as a true typical day). Keep in mind that this is not the number who enter the store, but the number who actually buy something.
2. Using that same research, determine the amount of money that each of these

typical customers on a typical day would spend in your store. For example:
20 customers × $50 per purchase = $1,000 sales per day

3. Decide how many days your store will be open in a year. Factors affecting this decision will mainly be the days that stores in your surrounding area are open, consumer shopping habits as they relate to your product line, and your own goals. Multiply the resulting dollar figure of sales per day by the number of days that your store will be open. This will determine the amount of business that might be generated in your store over the course of a year.

   In the preceding example, our ladies' clothing store has decided to be open 360 days per year (closed on Christmas Day, Thanksgiving, New Year's Day, Easter, and Fourth of July):

   $1,000/day × 360 days per year = $360,000 per year

   Note: When deciding upon the days and hours of operation of your store, take into consideration the opening and closing times of your neighboring stores, particularly if your store will depend upon activity in other stores to help draw customers into your store. You could end up wasting money on payroll and utilities if you stay open past the time that customers typically shop in your area. Another consideration is safety; you would not want your employees to be subject to danger that might exist after other stores have closed for the day.

4. Your calculations are not complete. You must now take into consideration your estimated sales per square foot. Using the research for similar stores that carry the same type of product, test your calculated sales above against this standard. In our example of a ladies' clothing boutique, our calculated sales plan was $360,000. Our store has a total of 3,000 square feet, 800 square feet of which is taken up with a stockroom, bathroom, and office space. We don't count our dressing rooms as we plan to install mirrors in those, thus making them part of our selling space. Subtracting the 800 square feet of non-

selling space will give us a net result of 2,200 square feet of selling space. If we divide the sales plan by the selling square footage of our store, the resulting sales per square foot would be $164 (rounded off).

$360,000 ÷ 2200 selling square feet = $164 per square foot

Fine-tune the Plan

We now have some soul searching to do. A quick look at the sales/square foot data (year 2002 to 2003) at BizStats.com will show that ladies' clothing stores will range from a low of about $110 gross sales/square foot for Cato stores (average store with 4,000 total square feet) to a high of about $849/square foot (selling space, average store size 2,306 feet) for Chico's. Of course, these numbers represent national averages and are achieved by stores that own units all over the United States, most of which have been in business for several years. Hopefully, you have also been able to find some information about sales per square foot results for small stores in your area. You will now want to take all of this information and draw some conclusions for your own store.

Using our ladies' clothing boutique example, we now want to analyze the numbers that we have tabulated thus far. We then think about the fact that our store is a new boutique that will have to wait until customers know and accept us before we experience optimum results. We also must take into consideration the population of our community (city), the health of the retail business in our selected area (information gleaned from our research and regional market feasibility study), and our specific product line and target customer profile. In all cases, it is strongly recommended that the student or new entrepreneur plan conservatively. It is better to plan low and revise sales estimates up as the business grows, rather than plan too high for the opening and risk overbuying inventory that may have to be sold at big discounts in order to move it out of the store.

With all of this in mind, we evaluate the situation and decide to take a conservative approach. Instead of $164 per square foot, we will plan our store at $150 per square foot for a resulting first year's sales plan of $330,000:

2,200 selling square feet × $150 per square foot = $330,000

RULE OF THUMB

For first-time store owners, it is best to keep your sales per square foot plan for most types of clothing stores between $125 to $250 per square foot.

TABLE 3.1 FIRST-YEAR SALES PLAN

Store location: 3445 West Cary Street	
Estimated buying customers served each day	20
Average price of items in store	$50
Dollar value of each day's sales	$1,000
Number days open each year	360
Amount of business per year	$360,000
Sales per square foot	
Total square footage	3,000
Selling square footage	2,200
Amount of business/selling square footage $360,000/2,200	$164 per square foot
Revised sales plan: 2,200 sq. ft. x $150/ sq. ft.	$330,000

Justifications:

1. The average price of items in the store was based on the weighted average price of each of our product categories. These retail price points were determined through research of key competitors and key brands that we plan to offer.
2. Through interviews with similar store owners and observations of sales activity in competing stores, we determined that approximately 20 customers would make a purchase on a "typical" day. Those customers would spend a total of approximately $50 on each visit to our store.
3. Based on the habits of other stores in our surrounding area, we decided to keep our store open 360 days per year.
4. Our first calculation yielded an annual first-year sales plan of $360,000.
5. A check of projected sales per square feet (selling square feet) showed a plan of $164 per square foot.
6. After reviewing these numbers, we elected to take a more conservative approach to our first-year sales plan. We decided upon a plan of $150 per square foot, yielding a final sales plan of $330,000.

A resounding yes is the answer (in almost all cases) to the question: "Is one million dollars as a sales plan for the first year of business too much?"

All financial documents need explanations or, as they are sometimes called, "justifications." Starting with the sales plan, you need to explain how you arrived at your projections and to "justify" the resulting figures. This does not require a dissertation; a few simple points combined with a couple of formulas will do just fine (see Table 3.1). See the Blank Financial Documents on your CD-ROM for a blank version of this table.

Calculating Fixed Expenses

The definition of ***fixed expenses*** or ***fixed costs*** are those monthly costs that will (roughly) be about the same every month and that must be paid regardless of the amount of sales you produce that month. For example, rent will be the same every month, and utility bills must be paid monthly. As for personnel costs, you *could* fire some employees if you anticipate a bad month, but that would be a very unwise business practice; especially considering that you will probably need them back again the next month. Since fixed expenses or fixed costs are the costs of running your business on a day-to-day basis, you need to accurately account for all of them before you start preparation of your financial documents. Other names used for these expenses are SG & A (selling, general, and administrative), overhead, and operating expenses.

Typical Fixed Expenses

In the simulation exercises at the end of Step 2, you completed a Store Location Fact List (Table 2.2), which included information about some of the fixed costs that you will have to pay each month as part of the operation of your business.

You now need to compile a complete list of fixed expenses. Table 3.2 gives examples of some of the common operating expenses that a small boutique will likely incur. This is the list of expenses that our simulated store would be obligated to pay each month.

An analysis of our fixed expenses shows the following:

- *Rental fees.* These are stipulated in the lease and would include any ***common-area fees*** and adjustments for sales growth, if there are any in the lease. Our rent is high in relationship to our sales plan. Ideally, it should be about 8 to 10 percent of our plan, but we are locating in Carytown, a trendy area of Richmond, and we feel that the extra expense for rent will pay off as we grow our business.
- *Staffing (also called payroll).* We have kept our staffing costs at 8 percent; there are two partners to also work the floor in the store, and although it is a large store at 3,000 square feet, we feel that this budget should be adequate. We have not included our salaries. According to Leroy A. Keller, Lender Relations Specialist for the Richmond office of the Small Business Administration, owner salaries should not be listed as fixed expenses. We will budget those later when we complete the cash flow plan.
- *Payroll expenses.* Just below staffing costs on the profit and loss (P & L) statement will appear a line for payroll taxes. These are federally mandated and include those costs that must be paid by the owner for each employee. These include ***FICA*** (Federal Insurance Contributions Act) charges for Social Security and Medicare coverage. You must budget 6.2 percent of payroll costs to cover Social Security and 1.45 percent for Medicare. In addition, you must also budget 6.2 percent of payroll for ***FUTA*** (Federal Unemployment Tax Act, which is unemployment insurance). We have rounded off the total of these expenses to 14 percent of total payroll expenses. Employees will also have money deducted from their paychecks for their share of Medicare and Social Security.

TABLE 3.2 FIXED COSTS (completed)		
COST CATEGORY	**ANNUAL EXPENSES**	**% OF SALES PLAN**
Staffing (payroll or wages)	$26,200	8.0%
Payroll taxes	$3,700	1.1%
Rent	$51,000	15.0%
Utilities (fuel, electric, water)	$5,220	1.6%
Maintenance and repairs	$1,200	.4%
Security	$1,080	.3%
Telephone/fax	$2,100	.6%
Internet	$480	.15%
POS system maintenance/ software updates	$1,200	.36%
Insurance	$2,080	.6%
Supplies and postage	$2,400	.7%
Marketing/advertising	$13,200	4.0%
Travel and entertainment	$5,000	1.5%
Accounting/bookkeeping	$4,200	1.3%
Banking services	$2,400	.7%
Miscellaneous	$2,400	.7%
TOTAL EXPENSES	**$123,860**	**37.5% of sales revenue**

- *Utilities.* Utilities to be paid by the renter include electricity, gas or fuel oil, and water. Utility expenses will vary from store to store and in different cities. Your real estate agent can assist you with projections for these costs, as he or she will have records of the facility's previous expenses (unless, of course, it is newly built). You may also contact the various utility companies for estimates. If you have difficulty getting this information, a neighboring friendly store owner might assist you by giving you good estimates of monthly utility costs. You may elect to list each utility expense separately on the list of fixed costs and the P & L, or you may combine them on the P & L. We have chosen to list costs for electricity separately and to combine fuel oil costs and water bills.

- *Maintenance and repairs budget.* This covers repairs to be made to equipment that breaks down or for any event when you need to call in a paid service person to fix something in the store.
- *Security.* This expense covers the monitoring fee if your store has a security system or if you plan to install one.
- *Telephone/fax.* We have combined these on one line because the total fees are low in comparison to other fixed costs. This expense includes both land and cell phone lines' monthly use costs plus fax expenses, if applicable.
- *Internet use and planned software updates.* If applicable, these make up the next two entries on the list.
- *Insurance requirements.* These will also vary by location and will usually be detailed in the lease. The decision about what kind of insurance to carry and how much coverage to purchase should be evaluated carefully: underinsure and you set yourself up for lawsuits, but if you overinsure, you risk eroding profits unnecessarily. Typical insurance for small stores includes fire, theft, and liability for injuries to customers and employees. For our simulated store, we contacted a Richmond insurance agent who has experience with small businesses in the area; in fact, we first met him at a ***SCORE*** (Service Corps of Retired Executives) workshop where he delivered a talk to small business owners on insurance needs. According to Jeffrey D. Virgin of Straus, Itzkowitz, & LeCompte, typically, a small store owner will need to carry the following kinds of insurance, and he also provides examples of allotment of money:

  1. Worker's compensation (laws differ by state) will cover us for medical bills for work-related injuries of employees, a portion of lost wages, and protection for us against lawsuits brought by employees and their families. We are required to have this insurance because we have three or more employees (including owners). This insurance will cost us approximately $1,425 per year.

2. Insurance for other liabilities can be covered under an umbrella policy, of which we have several choices depending upon desired coverage. We chose a less expensive (but not the least expensive) plan, which will cost us about $1,150 per year. This will cover us for such emergencies as liability for personal injury to customers and others who may get injured on our premises (up to $2 million), theft, damage to personal vehicles or rental vehicles when used for business, and loss due to fire.
3. Our lease stipulates that we must add extra insurance coverage for the landlord's property. This will cost an additional $25 per year.
4. Jeffrey D. Virgin based his estimates on information that we provided to him about our business: (a) an estimated annual payroll of $80,000 including owner salaries and (b) estimated value of inventory and fixtures/furniture inside the store of $60,000 (policy has a built-in allowance for an increase to value of inventory of +25 percent at peak selling times).
5. He explained that our total of $2,600 annual payments would be allotted in this way: a down payment (start-up costs) of $520 plus eight monthly payments of $260 per month. We will therefore calculate $520 into the capital expenditure list to start our business (Step 10) and the balance of $2,080 will appear on our fixed-cost list.
6. If you decide to offer health insurance or if you need auto insurance for company cars, those costs will be extra. Check your lease for other required insurance coverage, such as flood insurance if your building is located in a flood zone.

- *Supplies and postage.* These expenses are combined on our fixed-costs chart. Supplies for our simulated store include cleaning supplies; office items like pens, copy paper, and printer ink; and wrapping paper and bows for gift wrap. Postage includes shipping costs for packages and stamps for mailing bills (unless you pay online). This does not include freight charges for merchandise that you receive into the store; such charges have already been calculated into the cost-of-goods figures. Shipping costs will probably be a small number, as you will likely not ship many packages. If you are returning defective goods to a supplier, you would charge the supplier for freight. If you are shipping goods to a customer who lives out of state, you would probably charge the customer for shipping costs. The exception might be that you would elect to pay shipping charges for a very good customer who typically spends a lot of money in your store; that decision is part of your marketing and customer service strategy. As for postage, the cost of sending out marketing pieces like direct mail would be budgeted as a marketing expense.

- *Marketing and advertising costs.* These are projected to be in the moderate cost range. Typically, the recommendation is 3 to 5 percent of the sales plan. Our simulation shows 4 percent.
- *Travel and entertainment.* This is where you budget your buying trips, any trend research trips, and entertainment that you might need to do as a cost of doing business. This budget will likely be low in the first few years and might need to grow as you expand your business. To find an accurate estimate for this line, you will need to decide where you and your partner will go to buy merchandise, how many times a year you will travel to these locations, and the estimated cost of each trip. Trade shows are held in many locations in key cities in the United States and abroad, and there are many wonderful places to travel to gather trend information both domestically and around the world. Although this all sounds exciting, it will not be in your budget to travel to the more exotic and expensive sites. Most small retailers buy their merchandise in the following ways:

    - Local and regional representatives of large branded manufacturing and wholesale companies may visit your store, or you may be able to meet with them at regional trade shows.

- You or your partner may visit trade shows that cater to your customer and merchandise. These are held in key cities around the country, and you should do some research to find the ones that are best for your product (e.g., performing an Internet search for fashion trade shows or consulting *Women's Wear Daily*). One of the best and most popular trade shows for men's, women's, and young fashions is the MAGIC Show in Las Vegas.
- Visits to showrooms during fashion weeks and market weeks in New York. Your key suppliers will welcome you to shop their new lines during the times designated as fashion weeks or market weeks for your lines of merchandise. Again, you will need to do some research on when market weeks for your products occur. Lingerie market weeks, for example, are different from sportswear market weeks.

Once you have identified your key suppliers, you might want to purchase goods from their Web sites, particularly for reorder or fill-in merchandise. We will discuss this further in Step 6 on merchandising, but for the purpose of your fixed-costs list, you will need to calculate the following:

- How many trips will you need to make each year to what locations?
- Will both partners travel to all locations or will you take turns traveling?
- Estimate the cost of each visit to each location. For example, if you travel to New York City, will you fly and what will that cost? Where will you stay and what will that cost? Don't forget taxes in this estimate, as they add quite a bit to the cost of a hotel room in a city like New York. Estimate the costs of meals, taxis/subways/buses to get around while you are there. If there is an expense for attending a trade show, add that in as well.
- Will you need to budget for any entertaining? When you are with a vendor, he or she will likely treat you to lunch or dinner, but some store owners like to reciprocate occasionally by buying their key vendors a

lunch or dinner. Locally, you might want to entertain people of influence such as the president of the Country Club or Women's Association in order to publicize your business. Again, you cannot afford to do this on a regular basis, but you should allow for some expense if you feel that it will benefit your business.

Once you have totaled these figures, include them in your list of fixed costs. Use Table 3.3, which is also featured on the CD-ROM, to enter these expenses.

Additional Fixed Costs

According to L. Andy Keller, Lender Relations Specialist for the Richmond, VA, SBA, only monthly expenses of $100 or more should have a separate listing on the P & L statement. Look at the example statement in this step, and you will see that although we chose to list Internet costs and security system monitoring separately on our fixed-cost list, we combined them on our P & L. You can choose to do this, or to combine small monthly expenses under your miscellaneous cost projections. In your justifications, you should explain which regular expenses will fall under miscellaneous and what portion of the miscellaneous budget will be allotted for emergencies and unexpected expenses.

You will need to budget for accounting/bookkeeping to aid in the financial control of your business. Even if you or your partner is trained in and very competent at accounting, you will at least need an ***accountant*** to prepare your taxes and to monitor your books on some kind of regular basis. There is a great deal of liability and potential pitfalls in operating a business, and you will want outside professional assistance to make sure that you are following all the rules and completing all required forms and documents as demanded. You and your partner will need to decide how much and what kind of help you will need here.

Will you want a ***bookkeeper*** on a regular basis, or will you depend upon one of you to manage a computer program for your financial transactions? Will

TABLE 3.3 FIXED COSTS (blank)		
COST CATEGORY	**ANNUAL EXPENSES**	**% OF SALES PLAN**
Staffing (payroll or wages)		
Payroll taxes		
Rent		
Utilities (fuel, electric, water)		
Maintenance and repairs		
Security		
Telephone/fax		
Internet		
POS system maintenance/ software updates		
Insurance		
Supplies and postage		
Marketing/advertising		
Travel and entertainment		
Accounting/bookkeeping		
Banking services		
Miscellaneous		
TOTAL EXPENSES		

you use a ***payroll service*** to issue payroll checks? If you retain an accountant for all your financial record keeping, you will still need to pay that person more at tax preparation time.

The line for banking services includes the cost of maintaining a business account at your bank, costs for cashier's checks or money orders for any vendors who require these for payment of merchandise, and costs charged by the credit card processing companies for Visa, MasterCard, and American Express transactions. You might also want to pay for a service like Tele-charge, which immediately verifies the status of checks that customers write in your store.

In our list of fixed expenses, we include a line for POS (point-of-sale) system maintenance and software updates. We are assuming that we will want a computer/cash register system that will handle transactions, keep inventory, and help to plan for purchases. This type of system is discussed further in Step 10 when we detail capital expenditures (start-up costs) for our new store. This is an ongoing expense for keeping the system running properly and for any new software updates that we may need.

Your business may have other fixed expenses that we have not included here, such as delivery charges, business consultants, ***commissions*** for salespeople, dues for national organizations that benefit your sales effort, publications, equipment rental, shipping expenses to send purchased goods to best customers, copy machine or other technology rental expenses, alteration expenses, gift wrapping and gift boxes, or others. Perhaps you want to know if you can include such items as car expenses in your list of fixed expenses. Consult with your accountant on these types of expenses, as the IRS has rather strict guidelines about what expenses can be tax-deductible and which ones cannot. Your accountant can determine whether you should list car expenses as business expenses (with the plan of using them as tax-deductible items). You should add to the fixed-costs list any additional expenses that you anticipate needing to pay on a regular monthly basis.

Important: After you tabulate your list, you must write ***justifications*** for your expenses. These are explanations for why you have included the items and where you obtained the information for the budget. These do not have to be long, nor do they need to be repeated for other documents once you have justified them. Remember that expenses that might seem obvious to you might not be instantly understandable to someone who does not know your industry or your particular business. For examples of justifications, see the sample student business plan in the Appendix.

What you will *not* include on your fixed expenses are the costs of items or services that you will purchase for your start-up but that will not need to be replaced for a number of years. These costs are called capital expenses and will appear on your cash flow statement and are discussed in Step 10.

Controlling Fixed Expenses to Generate Profit

We cannot overstress the importance of controlling your fixed expenses from the very beginning, starting with the rental costs of your site. As stated in Step 2, you ideally would like to keep those expenses at or under 10 percent of annual sales. Start your business with only those expenses that you deem necessary and then increase the costs as your business grows, if you find it prudent. As with all matters, however, you want to strike a balance as much as possible: don't underbudget your fixed costs to the point where you are not providing the right atmosphere, staffing, and services for your product line and target customer. Also, plan realistically. Don't underestimate expenses just so that it will look good on paper. You don't want any surprises later when you find that an expense is actually much higher than you had budgeted.

Accountants and Lawyers

In the discussion of fixed expenses, we talked a little about the importance of allocating funds for the regular services of an accountant or bookkeeper. Carol Amernick, CPAPC and owner/manager of Amernick accounting firm in Richmond, Virginia, specializes in working with certain types of small business owners. According to Carol, the potential store owner should consult first with a good accountant/CPA, second with an attorney, and third with a banker. "The smaller your operation," she says, "the more you need this kind of help because any misstep can cause you to lose everything."

Accounting Services and Costs

An accountant can perform several services for the entrepreneur. First, the accountant can help you with the completion of, or at least review of, the financial documents for the business plan. Even though you are trying to save money, don't overlook the value of having an accountant review your plans. Second, the accountant can advise and assist in the selection of a system to manage your bookkeeping and stock-keeping efforts. Certainly, computer programs are available for this purpose, such as QuickBooks, MSN Money, or Peachtree, and a good accountant can direct the entrepreneur in which of these programs (or others) are best. Third, an accountant or bookkeeper can control payroll expenses, write payroll checks for employees and owners, and manage payroll taxes and benefits. Fourth, you will certainly want an accountant who understands small business practices to prepare your year-end tax returns. Included with this service is advice on how you can minimize tax expenses.

What will these services cost you? If you intend to use the accountant or bookkeeper on an ongoing basis, you can probably negotiate a "package deal" for weekly, bimonthly, or monthly help with the business operation that will include the preparation of year-end taxes. Most accountants who work with small businesses charge about $200 to $250 per month if they also do payroll; year-end tax preparation will likely run an additional $600 to $1,200, depending upon complexity and hours spent in preparation. If you just want payroll checks to be written by a service firm, that will likely cost between $25 to $50 per month. Charges vary by area, so you will need to do some research in your locale.

You also might wonder at this point where to find a good accountant or bookkeeper. As with many other questions such as this, word of mouth with good recommendations from other business owners is the best source.

Find a Good Lawyer and Budget the Cost

What about lawyers? As stated earlier, you probably will not need to keep a lawyer on ***retainer*** for your small business, but you certainly need to establish a working relationship with one who understands and services small businesses (some large firms do not work with small owners, and larger firms tend to charge higher rates). As with the accountant, word of mouth is your best source for finding a reputable, competent lawyer who understands small business operation, but you may also contact the bar association in your area for a list of names and specialties or ask your accountant for a referral.

You will want the help of a lawyer when you are setting up your business: to advise you on business structure and form of ownership, read your lease and recommend changes if needed, advise on employee issues, and confirm types of insurance that you will need. Carol Amernick estimates that the cost of basic start-up paperwork from your lawyer will be from $600 to $1,200. Again, you will need to research costs in your own locale. Remember that these start-up costs will be part of capital expenses, which are covered in Step 10.

Establishing Goals for the Business

In the Conclusion of this book, we go into more detail about how to handle the successes and failures of your business, but before you start your financial planning for the opening, it is good to take a few minutes to think about long-range goals. All effective managers have an ongoing plan: a one-year, two-year, five-year, and possibly even a 10-year plan. Thinking and planning ahead will help you to make important decisions about what you do with your business today. For example, if you think that you want to stay in a small location for two to three years until you get the business going strong and then relocate to larger or more expensive quarters, you might take a different approach to investing in fixtures and renovations to the space than if you plan to stay in your first location for five years or more. If you decide that you and your partner will work 24/7 for the first couple of years to save personnel costs, but you want to be able to pursue other endeavors (like opening a second store) a few years down the road, then you will want to hire personnel more carefully with the thought of finding someone whom you can promote to manager in a few years.

In any event, you should make a five-year plan for yourself and your business, with the understanding that you can and should revise this plan as you grow in your business. Even though your financial documents for the plan will only represent three years initially, you will want to have an outline of what you hope to accomplish in five years. This will help you plan finances, personnel, location, and marketing of your business.

For the short term, take a realistic approach to all aspects of your business plan for the first two years especially. Know your limitations financially and personally. Yes, you will work hard, but don't overextend yourself or your partner to such an extent that you either fall ill or can't live up to your obligations to the business or perform the necessary planning and monitoring. Also, be realistic about the amount of time that it will take to achieve big sales successes. There are very few overnight success stories. Most take a lot of time, effort, and even luck.

Simulation

1. Using the guidelines discussed in this step, develop your sales plan for year one of your business. Justify that plan with an explanation of methods and sources used for formulation of that annual sales estimate: observation of customer activity, interviews with store owners, research on typical new store results, and industry standards. Using that first estimate, determine sales per square foot (selling space) for your store and compare that number to industry standards, results of other stores in your area, and so on. Finally, use your research and best judgment to decide on a definite sales plan for the first year of business. Justify that decision. After you have fine-tuned this document, it will become part of your Part IX Financial Data in your final business plan.
2. Starting with the information that you filled in when you completed Table

3.3, make a list of fixed expenses that your store will have to pay each month. Justify those expenses through your research and interviews, and give an explanation for each expense, including notes on how you intend to control fixed expenses in order to be profitable. This list should be researched thoroughly for accuracy, as it will become an important part of the P & L plan that we will execute in Step 5.

3. Make a short list of long-term goals for your business. You will want to modify or add to this list after evaluating the information in Step 10.
4. Begin to research types of investors and places where you can find money to start your business. You will need to detail this information when we prepare our cash flow statement in Step 10.
5. Refer to the completed business plan in the Appendix to assist you with the completion of the sales plan and the list of fixed costs. Both appear in the Financial section of the plan.

STEP 4
Preparing a Business Resume and Personal Financial Statement

"What you see is what you get."

PROJECT OBJECTIVES

This step in the simulation will guide you through:

- Guidelines for preparing a business resume
- Differences between business and personal resumes
- How to complete a personal financial statement

The Business Resume

In most instances your business resume will be your first opportunity to show who and what you, as a businessperson, are all about. In developing the resume, you must have a short, concise, and accurate document that tells an interested party that what they see on the paper in front of them is what they are going to get. That interested party most likely will be someone who may want to invest in your business, loan the business money, work for the business, or buy what the business is selling.

Remember too that it is extremely important when you are promoting yourself as a businessperson that you act the part. If you do not take yourself seriously, no one else will either. To that end, having a well-designed business card that you can present is a prime indicator that you are indeed someone who takes his or her business seriously. In addition, you need to have stationery with a professionally printed business letterhead and include the business logo on all of your e-mail correspondence. (See our stationery and business cards in the Rose Knows Clothes simulated store folder on the CD-ROM.)

The business resume therefore needs to include not only the normal kinds of things one sees on a resume (name, location, contact numbers, etc.) but it also needs to include the following:

- Business experience of the owner
- Mission statement of the business
- Detailed descriptions of the products or services that the business sells
- Type of market that the business hopes to attract
- Special expertise that the owner or employees may have
- *Geographic footprint* of the business
- Important relationships that the business may have with suppliers, manufacturers, or consultants
- What is unique about the business
- Web site of the business
- Educational background of the owners

There is no one-size-fits-all business resume, but the preceding items should be part of it. In reality, the business resume is the business plan summarized in one or two pages. For a sample of a business resume, review the sample business plan in Appendix.

Difference between a Business and Personal Resume

You might think that all resumes are the same. Well, yes and no. As we have already determined, a resume is a summary of experience, so they do have common elements. However, the type of resume depends on what kind of experience you are summarizing. An obvious assumption is that a personal resume is a summary of personal experience—that is, education, jobs held, honors received, and so on. Just as obvious is that the business resume is a summary of the business experience of the owners.

While the two different resumes have a similar goal in introducing their subject, the type of information provided will be based on the objective of the person/organization preparing it. The main difference is that the goal of the personal resume is to introduce a person to a business—generally to get that person a job as an employee of the business. On the other hand, the goal of the business resume is to introduce the business and its owners to a multitude of constituencies—the investor, the banker, the prospective employee, and ultimately, the customer. So it is easy to see that the personal resume is focused on one target—the potential employer—while the business resume has many targets as its focus.

Personal Financial Statement

Lenders and investors in your business will want to know how good you are at managing not only the business but your personal finances as well. Since the money that they are providing to you must be paid back, the lender will want to know the borrower's personal financial status. This information is especially important when the borrower is seeking an unsecured loan because the lender will want to know the financial worthiness of the person who is responsible for paying it back.

When the borrower submits an application for a loan, the lender will in almost all cases ask the borrower for a ***personal financial statement (PFS)***. The form that is provided will vary from lender to lender, but it always asks for the same information. When completed, it gives both the lender and the borrower a thorough picture of the borrower's personal financial well-being.

As one might expect, completion of the PFS can be easy or difficult depending on the complexity of the borrower's personal financial situation. However, it is vital that all of the information on the PFS not only be entirely accurate but timely as well. Most lenders will require that the data on the PFS be less than 90 days old, so it must be continually updated.

Elements of the Personal Financial Statement

The common PFS has the following characteristics and requests for data:

Section I—Borrower Information

Name: Name of the borrower whose data is being provided.

Address: Current address of the above-named person.

Phone number: Home and business phone of the above-named person.

Business: Business name of the borrower.

Section II—Assets

Cash on hand and in banks: Total amount of cash owned by the borrower in their possession and in banks.

Savings accounts: Total amount the borrower has in all savings accounts.

Retirement accounts: Total amount the borrower has in all retirement accounts.

Accounts receivable: Total amount the borrower is owed by others.

Life insurance: Cash value of all life insurance policies owned by the borrower. Include the name of the insurance company and the names of beneficiaries.

Stocks and bonds: Current market value of all stocks and bonds owned by the borrower. List the names of the securities, their cost, and the date of quote of the current market value.

Real estate: Current market value of all real estate owned by the borrower.

Automobile: Present value of any vehicles owned by the borrower.

Other personal property: Value of any other personal property owned by the borrower (jewelry, paintings, boat, etc.) and any lien information.

Other assets: Value of any assets not already included.

Total: Value of all assets owned by the borrower.

Section III—Liabilities

Accounts payable: Amount owed by borrower on accounts (e.g., credit cards).

Notes payable: Amounts owed by borrower to banks and others. Include the original balance, current balance, payment amount, payment frequency, and how secured (signature, collateral, etc.).

Installment account (auto): Amount owed on auto loan. Include the amount of the monthly payment.

Installment account (other): Amount owed on other installment loans. Include the monthly payment.

Loan on life insurance: Amount owed on loans on the borrower's life insurance.

Mortgages on real estate: Amount owed on any real estate mortgages. Include the address of the property as well as the date it was purchased, its cost, and mortgage information (name of mortgage holder, balance, and payment schedule).

Unpaid taxes: Amount owed by borrower in unpaid taxes.

Other liabilities: Amount of borrower's liabilities not already listed

Total liabilities: Total amount owed by the borrower.

Net worth: Subtract total amount owed by the borrower from the total value of all of the owner's assets.

continued

Section IV—Sources of Income and Contingent Liabilities

Most PFSs require the borrower to list sources of income and contingent liabilities in order for the lender to further assess the financial health of the borrower. Sources of income include the salary of the borrower, investment income, and real estate income. Contingent liabilities include the borrower being an endorser or co-maker on a note and legal claims or judgments.

Sample Personal Financial Statement

Table 4.1 is a completed sample personal financial statement utilizing the format developed by SCORE. SCORE is a volunteer organization funded and promoted by the U.S. SBA, whose mission is to assist those wishing to start their own businesses. The reader can also access a blank form on the CD-ROM in the Blank Financial Documents folder. There are also further instructions for completing it at www.score.org.

TABLE 4.1 PERSONAL FINANCIAL STATEMENT (completed)

Rose RagstoRiches Date:

ASSETS	AMOUNTS IN DOLLARS
Cash: checking accounts	4,000
Cash: savings accounts	9,000
Certificates of deposit	10,000
Securities: stocks/bonds/mutual funds Notes and contracts receivable	5,000
Life insurance (cash surrender value)	8,000
Personal property (autos, jewelry, etc.)	20,000
Retirement funds (e.g., IRAs, 401k)	–
Real estate (market value)	200,000
Other assets (specify)	1,000
Other assets (specify)	7,000
TOTAL ASSETS	**$264,000**

continued

TABLE 4.1 *(continued)*	
LIABILITIES	**AMOUNTS IN DOLLARS**
Current debt (credit cards, accounts)	5,000
Notes payable (describe below)	10,000
Taxes payable	–
Real estate mortgages (describe)	175,000
Other liabilities (specify)	–
Other liabilities (specify)	–
TOTAL LIABILITIES	**$190,000**
NET WORTH	**$74,000**

DETAILS 1. Assets						
Notes and Contracts Held						
LENDER	**BALANCE OWING**	**ORIGINAL AMOUNT**	**ORIGINAL DATE**	**MONTHLY PAYMENT**	**MATURITY DATE**	**PURPOSE**
Brother	$5,000	$7,000	10/1/2007	$100	12/1/2011	College loan

Securities: Stocks/Bonds/Mutual Funds				
NAME OF SECURITY	**NUMBER OF SHARES**	**COST**	**MARKET VALUE**	**DATE OF ACQUISITION**

continued

DETAILS *(continued)*
1. Assets

Stock in Privately Held Companies

COMPANY NAME	NO. OF SHARES	$ INVESTED	EST. MARKET VALUE

Real Estate

DESCRIPTION/ LOCATION	MARKET VALUE	AMOUNT OWING	ORIGINAL COST	PURCHASE DATE
Residence/ 123 VCU Street	$200,000	$175,000	$200,000	8/1/2008

2. Liabilities

Credit Card and Charge Card Debt

NAME OF CARD/CREDITOR	AMOUNT DUE
VISA	$2,500
MasterCard	$2,500

Notes Payable (excluding monthly bills)

NAME OF CREDITOR	AMOUNT OWING	ORIGINAL AMOUNT	MONTHLY PAYMENT	INTEREST RATE	SECURED BY (LIEN)
Auto Dealer	$10,000	$20,000	$200	5.00%	Lien on Auto

continued

DETAILS *(continued)* 2. Liabilities					
Mortgage/Real Estate Loans Payable					
NAME OF CREDITOR	**AMOUNT OWING**	**ORIGINAL AMOUNT**	**MONTHLY PAYMENT**	**INTEREST RATE**	**SECURED BY**
Parents	$175,000	$193,000	$1,500	0.00%	Residence

Simulation

1. Each partner in the business must prepare a business resume. This form will be included at the end of the Personnel section of the business plan.
2. Each partner in the business must prepare a personal financial statement. These statements will be included at the end of the Personnel section of the business plan.

STEP 5
Profit and Loss Plan

"Beware of little expenses; a small leak will sink a great ship." —*Benjamin Franklin*

PROJECT OBJECTIVES

This step in the simulation will guide you through:

- Calculating a realistic store gross margin plan
- Understanding key formulas that will help you monitor your success in your new business, and how to use them
- Extending your annual sales plan into a monthly sales plan that works for your product line
- Calculating a breakeven analysis
- Preparing Profit and Loss plans

The simulation instructions at the end of this step require you to begin the serious business of putting together the financial documents that will allow potential investors—as well as you and your partner(s)—to see if you have the necessary understanding, ability, and concept to run a successful (i.e., profitable) store. We will continue to emphasize (ad nauseam) throughout this book the necessity of profit in any store operation. You might have the most creative, novel idea for a store and product line that has ever graced the retail landscape, but if you can't make money running that business, you might as well stay home or take your money to Vegas, where you will likely have the same odds of holding on to it! With this thinking in mind, we need to take the necessary steps to preparing a realistic profit and loss plan—one that takes into account the fixed monthly expenses that must be paid regardless of sales activity as well as sales and gross margin goals that are achievable for your kind of business. It should be noted at this point that making a profit requires planning, work, and a certain amount of good luck. As mentioned, you may only realize a small profit the first year of your business, and that is not atypical, but it is crucial that you have the tools and take the appropriate amount of time to monitor your business regularly to be prepared to make needed adjustments to get your store into a profitable position and be able to maintain it.

In Step 3, you were asked to prepare a list of fixed expenses for your business and estimate a sales plan for your first year of business. This step includes a Gross Margin Exercise that will help you plan the gross margin percent for your first year of business.

Gross margin, also called gross profit, is the amount of money that you will have left over from the total sales that come through your cash register after you pay for the cost of your merchandise (in our simulation, that equates to cost of the product plus the cost of freight to get it to your store). Workroom/alteration costs as well as any cash discounts earned on invoices will also impact your ***cost of goods sold (COGS)***. It is crucial to get a realistic fix on this projected figure, as you will need to pay all your fixed expenses from the gross profit dollars, and the balance of any funds left after paying fixed expenses will need to cover income taxes and (hopefully) still give you some money to give back to the partners and/or reinvest in the business. It is important not to be too optimistic when determining your gross profit dollars and your gross margin percent; you don't want any big surprises when you look at profits at the end of year one in business.

Gross margin (or gross profit) can be expressed as both dollars and percent; for this next exercise, our goal is to calculate a planned gross margin percent for year one that we can then plug into our projected profit and loss plan. With this calculated percent, we can then proceed to complete the profit and loss projections for year one.

Calculating Annual Gross Margin Percent

To project our gross margin percent for planning purposes, we must first make some assumptions. An entrepreneur might pull together his or her knowledge, experience, research, and gut feeling about how planned retail price points and anticipated markdown percents will result in a gross margin percent for a new store. Table 5.1 on page 94 shows an example of this.

Retail Selling Prices

When a buyer or store owner makes a wholesale purchase from a vendor, the vendor will charge him a certain price for the merchandise. To this price must be added the cost to ship that merchandise from the vendor's warehouse to the retail store's location. Unless the buyer is part of a huge multistore chain with mega buying power, the buyer/owner must pay the cost of that freight. On a profit and loss statement (which we will complete later in this step), freight to get said merchandise from the vendor warehouse to the store location is sometimes listed as a separate fixed-cost line below the gross profit line. For the entrepreneur, you may wish to list your freight costs in that manner. For students, we are going to simplify the P & L statement by assuming that the cost of freight is included in the COGS prices.

When the vendor's initial cost is added to the freight cost, the buyer/owner calculates an *initial* markup percent that she needs to achieve in order to make a profit on that item. An initial markup is the first markup, before any retail price reductions such as promotional markdowns or permanent markdowns are taken (see also Step 6 on merchandising).

In an ideal world, retailers would be able to sell all their merchandise at full retail price and the full initial markup, but alas, we in the world of fashion and retailing do not live in an ideal world. No matter how good a buyer/owner is at selecting merchandise, there will always be markdowns that need to be taken. During the season, stores usually have some promotional events such as anniversary sales, cooperative sales events between sister stores in an area, or a town event such as Richmond's Watermelon Festival every July. During these times, the owner will want to give the shopper an extra incentive to buy, such as a percentage off a group of merchandise, a category of goods, or even the store's entire stock. At the end of a selling season (summer, for example) or if a group or item of merchandise proves to be undesirable to the customer, a store owner will need to take permanent markdowns in order to clear inventory and make room for more saleable goods. We cover the fine points of taking markdowns in more detail in Step 6, but for now, it is important to understand that your annual sales dollar plan (gross sales) is equal to all sales (receipts) that come into the business and that markdown dollar sales are included in those gross sales. In other words, not everything that you sell will be at regular (full) price.

Calculating Markup

Let's talk math for just a bit. The formula for calculating markup is the following:

$$\frac{\text{Retail \$} - \text{Cost \$}}{\text{Retail \$}} = \text{Markup \%}$$

Example:

$$\frac{\text{\$24 retail price} - \text{\$10 cost price}}{\text{\$24 retail price}} = \text{58.3\% markup}$$

This formula is great and very useful, but what happens if you know the retail price that you want to charge and you know your markup percent goal? How then can you determine the cost that you can afford to pay for the item? Easy; we use a number called the ***reciprocal*** (or cost complement) of the markup to calculate our cost figure. The reciprocal is a handy number that is quite simply the difference between 100 percent and our markup goal percent (in this example, 58.3 percent). Remember that the markup percent and the reciprocal percent must add up to 100 percent. So, if we know that we want to retail a particular item for $24 and we know that our markup goal is 58.3 percent, we use the reciprocal (100% – 58.3% = 41.7%) to calculate our cost price:

$24 retail price × .417 (reciprocal percent) = $10.00 cost.

Figuring Wholesale Costs

How, you may ask, can you know the cost prices of your proposed merchandise items? Entrepreneurs will be able to get this information as you research your industry and speak to the vendors with whom you plan to do business. Students may have access to this information if they currently or in the past have worked at a retail boutique and had access to cost of merchandise information. They might also be able to speak to vendors directly or interview cooperative retail boutique owners in their areas and ask questions about typical cost prices. Again, try to avoid direct questions about what they pay; instead ask, "On [insert certain types of merchandise], what are the 'typical' cost prices?" If students cannot find cost information on all their items or categories, they can use the handy reciprocal formula to calculate costs.

Guidelines for Students

Most small clothing boutiques operate on an initial markup on goods of 55 percent to 65 percent, depending upon item, category, competition, and importance to the consumer. Some stores use even higher initial markup percents, particularly if they do not have much competition in their areas or if it is a very "hot" fashion item. Price points also help to determine markup percents; luxury goods are typically sold at a much higher markup percent. In all cases, however, it is critical to be aware of two indisputable facts: First, your competitors will help to set your retail prices and, thus, your markup percents. You can't sell the same item for more than your competitors if you expect to get customers to shop in your store. Second, consumers will "vote" on whether or not they are willing to pay your prices. If they are too high, you will quickly know that by the absence of money in your cash register.

Sales and Gross Margin Assumptions

Now that we have set some ground rules, let's determine some assumptions about our product assortment, price points, and markup percents. In Step 6, you will have the opportunity to plan your merchandise opening assortment, but for this exercise, you need to make some assumptions based on prior knowledge, experience, and research.

As you begin to finalize your product categories, look at some successful small retail boutiques in your area and notice how many and what kinds of merchandise categories they seem to sell successfully. Here in Richmond, Virginia, we looked to several successful small stores to see what they offer, as shown in Figures 5.1 to 5.5 (See the store owner interview folder on the CD-ROM to learn more about these stores.)

Figure 5.1 *Pink in Richmond, Virginia, is a small women's boutique for the fashion-savvy college student and young career woman. They also carry an extensive handbag line.*

Figure 5.2 *Lex's of Carytown is a small specialty store that caters to females of all ages, from teens to women in their fifties.*

Figure 5.3 *The Heidi Story store in Richmond, Virginia, is a women's boutique with a twist. The owner teaches sewing classes to children and adults and does alterations as well as special orders.*

Figure 5.4 *Que Bella is a small boutique that sells higher quality shoes and handbags, including accessories.*

Notice that each of these stores has a particular niche, and that they do not try to be all things to all customers. After your research and some soul searching, you are now ready to finalize your own product categories, determine retail pricing and ***wholesale costs***, calculate both ***initial markup*** and ***maintained markup*** for each category, and, finally, determine overall anticipated gross margin percent for your store. You will need all of this information when you prepare your P & L statement.

Table 5.2 is a method for listing your assumptions and calculating some key

Figure 5.5 *Rumors a small specialty boutique for young, edgy customers. Its primary market is college women who are located right in the heart of the Virginia Commonwealth University academic campus.*

TABLE 5.1 SALES AND GROSS MARGIN EXERCISE ASSUMPTIONS (completed)
Annual first-year sales plan: $330,000

Part A: Category % of total annual planned sales

CATEGORY NAME	PLANNED PERCENT	PLANNED DOLLARS
Dresses and skirts	25%	$82,500
Bottoms: pants, shorts	20%	$66,000
Tops: blouses, sweaters, shirts	25%	$82,500
Jackets and outerwear	15%	$49,500
Accessories	15%	$49,500
Totals	**100%**	**$330,000**

Part B: By category, % of sales for regular price vs. markdown price

CATEGORY NAME	REGULAR-PRICED SALES %	REGULAR-PRICED SALES $	MARKDOWN SALES %	MARKDOWN SALES $
Dresses and skirts	65%	$53,625	35%	$28,875
Bottoms: pants, shorts	70%	$46,200	30%	$19,800
Tops: blouses, sweaters	70%	$57,750	30%	$24,750
Jackets and outerwear	60%	$29,700	40%	$19,800
Accessories	75%	$37,125	25%	$12,375
Totals		**$224,400**		**$105,600**

Part C: By category, average percent reduction for all markdown sales (promotional plus permanent)

CATEGORY NAME	AVERAGE REGULAR PRICED	AVERAGE % REDUCTION	AVERAGE MARKDOWN PRICE
Dresses/skirts	$129.00	-40%	$77.40
Bottoms	$113.00	-30%	$79.10
Tops	$65.00	-35%	$42.25
Jackets/outerwear	$167.00	-40%	$100.20
Accessories	$52.00	-25%	$39.00

numbers that will allow you to complete the Sales and Gross Margin Exercise later in this step. Use Table 5.2, which is also featured on the CD-ROM, to formulate your own plan. There are three parts to the assumptions:

Part A is where you list your merchandise categories. We explored these in Step 1, and now you need to decide how to group your merchandise into categories for the sake of planning. As mentioned, you should limit your categories to between three and five, and make them broad enough to allow flexibility for seasonal variation, fashion fluctuations, and changes in customer expectations. Notice in the example that the first category has both dresses and skirts. If dresses become less fashionable for a particular season, we can allot more purchase dollars to skirts. You should start this exercise by deciding what percentage of your total planned gross sales will be in each category. You can then calculate planned dollar sales for each of these categories.

Part B directs you to take the dollars by category that you have just calculated and to decide what percentage of each category will come from regular-priced (full-price) sales and which will come from markdown (either promotional or permanent) sales. How to determine this? We wish that there was one easy answer to all of this, but you must use your prior knowledge, research, and interviews to help you. You can look for articles on results of particular stores, consult certain Web sites like BizStats.com, check for information on the Web site for the National Retail Merchant's Association, ask local retailers and merchant's associations, and consult other sources for help with these figures. You should write down a plan for your store that includes times when you will promote your merchandise and which categories will be promoted. From your plan, you can determine which categories are going to be promoted the most often and which categories are the most "risky" in terms of fashion timing. If a category like dresses and skirts is likely to have more changes in fashion direction, you might be at risk to mark down more units in those categories than you would in accessories. (This depends upon what kind of accessories and how trendy they are.) Seasonality is a factor; jackets and outerwear typically carry a higher percentage of markdown goods because it is so hard, for example, to sell ski jackets in July. Use your best judgment and seek the advice of business professionals to help you with this part of the exercise. After you have calculated percentages for both regular-priced and markdown goods, you should calculate dollar sales for each of these, by category.

Part C takes the information one step further. For this exercise, you will need to determine average retail price points per category. To do this, start by researching retail price points of similar merchandise that you plan to carry; look for this merchandise in stores that are similar to yours and that solicit the same target customer. Decide upon four average price points per category, and then determine a "weighted" average retail price point for each category. In other words, decide at which price point you will do the most business and give that more weight in your calculations. You might write down your price points in this manner:

Dresses and skirts retail prices:

$100	35%
$125	40%
$160	15%
$200	10%
	100%

If you multiply each retail price point times its corresponding percentage and add those results together, you will get a weighted average retail for this entire category. In the case of this example, the answer is $129. Next, complete this exercise for every one of your product categories. You will need to keep this list and refer back to it when you do the Sales and Gross Margin Exercise later in this step, and then you will need the entire list with all your price points when you complete the Opening Inventory chart in Step 6 (Table 6.1).

TABLE 5.2 SALES AND GROSS MARGIN EXERCISE ASSUMPTIONS (blank)
Annual first-year sales plan

Part A: Category % of total annual planned sales

CATEGORY NAME	PLANNED PERCENT	PLANNED DOLLARS
Totals	100%	

Part B: By category, % of sales for regular price vs. markdown price

CATEGORY NAME	REGULAR-PRICED SALES %	REGULAR-PRICED SALES $	MARKDOWN SALES %	MARKDOWN SALES $
Totals				

Part C: By category, average percent reduction for all markdown sales (promotional plus permanent)

CATEGORY NAME	AVERAGE REGULAR PRICED	AVERAGE % REDUCTION	AVERAGE MARKDOWN PRICE

Note: You must complete this exercise for each of your categories and save the list, including all your retail price points with their corresponding weighted average retails. You will need this information later.

Next, decide what will be the average percentage that you will need to take off the regular price over the course of the year, in order to get to the average markdown retail price. Take the information from Part B, and then note on this plan how you will handle permanent and end-of-season markdowns. From your observation of other similar stores in your area, you should be able to determine, for example, a markdown plan for the dress and skirt category that is similar to the following:

- Keep new merchandise at regular price for the first four to six weeks after arrival in the store.
- After that, markdown any remaining stock to 20 percent off.
- After two more weeks, mark down any remaining pieces to 35 percent off.
- After two more weeks, mark down leftover pieces to 50 percent off.

TABLE 5.3 SALES AND GROSS MARGIN EXERCISE (completed)

CRITERIA	DRESSES	DRESSES MD	BOTTOMS	BOTTOMS MD	
Annual sales %	16%	9%	14%	6%	
Annual sales $	$53,625.00	$28,875.00	$46,200.00	$19,800.00	
Avg cost each	$55.76	$55.76	$49.85	$49.85	
Bags, wraps	$1.00	$1.00	$1.00	$1.00	
TOTAL COST EACH	**$56.76**	**$56.76**	**$50.85**	**$50.85**	
Avg retail each	$129.00	$77.4	$113.00	$79.10	
Gross profit $ each	$72.24	$20.64	$62.15	$28.25	
Gross profit % each	56.0%	26.7%	55.0%	35.7%	
Reciprocal %	44%	N/A	45%	N/A	
GROSS PROFIT	**$30,030.00**	**$7,700.00**	**$25,410.00**	**$7,068.60**	
GROSS MARGIN					

Note: Abbreviation MD stands for markdown retail price points at regular price are rounded to replicate typical store pricing.

◆ If there are any pieces left after the last markdown, you can sell them to a jobber, donate them to charity, or pull together an assortment of these items and have a sidewalk sale.

You need to analyze each category in this manner. Your plan does not have to look just like this, but you should formulate a plan.

With the listing of the set of assumptions, you are now ready to complete the Sales and Gross Margin Exercise. (See Table 5.3 for an example of a completed exercise.)

To complete Table 5.4 on page 100 (also on the CD-ROM), use your table of assumptions and follow these steps:

◆ Fill in the planned sales dollars by category, both regular-priced and markdown price sales into row 2, Annual Sales Dollars.

◆ Calculate percentages for row 1 by dividing each column of planned sales dollars by the total annual plan ($330,000 in the example).

◆ If you know the average cost of your planned merchandise by category, you

	TOPS	TOPS MD	JACKETS	JACKETS MD	ACCESSORIES	ACCESS MD	TOTAL
	18%	8%	9%	6%	11%	4%	100%
	$57,750.00	$24,750.00	$29,700.00	$19,800.00	$37,125.00	$12,375.00	$330,000.00
	$28.25	$28.25	$69.14	$69.14	$19.80	$19.80	
	$1.00	$1.00	$1.00	$1.00	$1.00	$1.00	
	$29.25	**$29.25**	**$70.14**	**$70.14**	**$20.80**	**$20.80**	
	$65.00	$42.25	$167.00	$100.20	$52.00	$39.00	
	$35.75	$13.00	$96.86	$30.06	$31.20	$18.20	
	55.0%	30.8%	58.0%	30.0%	60.0%	46.7%	
	45%	N/A	42%	N/A	40%	N/A	
	$31,762.50	**$7,623.00**	**$17,226.00**	**$5,940.00**	**$22,275.00**	**$5,779.13**	**$160,814.23**
							48.7%

may proceed next to fill in row 3. If you do not, then use the directions above for figuring cost from retail prices, using the reciprocal of the markup and the planned markup percentage by category. If you do not know what markup you should use, then use the guidelines listed on page 88. Keep in mind, however, that if you expect an initial markup percent of 60 percent, then you need to subtract 1 to 2 percent from that number as your goal, to allow for freight costs to get the merchandise to your store (depending upon how far your vendor warehouses are from your store location).

- In row 4, enter an estimated cost for shopping bags, tissue wrap, and any other supplies that you will use for every customer's purchase. If you plan to offer gift-wrapping services, enter those costs later as supplies. Since not every customer will need that service, the cost does not need to be listed here. Your bag and wrap costs will typically be $1.00 to $2.00 per item, depending upon how fancy you expect these items to be. You can check out typical costs online. If you use just a simple shopping bag and some white

TABLE 5.4 SALES AND GROSS MARGIN EXERCISE (blank)

CRITERIA		MD		MD	
Annual sales %					
Annual sales $					
Avg cost each					
Bags, wraps					
TOTAL COST EACH					
Avg retail each					
Gross profit $ each					
Gross profit % each					
Reciprocal %					
GROSS PROFIT					
GROSS MARGIN					

tissue paper, your cost will be closer to $1.00 each. If you intend to have gold-engraved bags of an atypical size with ribbon and colored tissue, your price would be higher. Certain stores, like bridal houses, might need to budget higher amounts to allow for things like zippered garment bags. Add that cost to the cost in row 3 and get a new total for row 5. From your assumptions in Table 5.2, enter the retail prices that you have determined for regular-priced and markdown-price goods. Use rows 7 and 8 to calculate gross dollars and percent per item. In row 9, you will use your gross profit percent and multiply it times the planned sales in row 2. Adding together the category gross profit dollars in row 9 and dividing that number by the total annual planned sales will give you the average annual planned gross margin for your store.

- To evaluate the final gross margin percent for your store against industry standards for your merchandise, you will have to do some research. See the Resources section at the end of the book for research options.

		MD		MD		MD	TOTAL

Note: We have used a number of abbreviations in this form to save space and get all the information on one page. You may abbreviate as you wish, but if any abbreviations are not readily understandable, add a note to the bottom of your form to explain them. You may also elect to spread your worksheet onto a bigger form, but make sure that it is all on one piece of paper to be reader-friendly. Note also that we have used both terms *gross profit* and *gross margin* on this form. These terms are interchangeable, and we use them both here just to differentiate between gross profit on a style and gross margin for the whole store.

Guidelines for Students

For most new apparel boutiques, your total gross margin percent plan for the store's first year should be conservative. Keep it under 50 percent. For both forms (Table 5.2 and Table 5.4) in your Sales and Gross Margin Exercise (as with all financial forms), you will need to provide a list of justifications that explains your calculations, along with any proof (research) to substantiate the numbers.

Monthly Sales Planning

Before we can proceed to complete the P & L statement, we need to take our annual sales plan and break it down by month. (See Table 5.5 for an example of a completed Monthly Sales Plan.)

To complete the form shown in Table 5.6 on page 104 (also on the CD-ROM), call upon your research and knowledge of your product lines and industry to determine the months when your sales will be average, higher than average, and lower than average. You can use the list of resources provided at the end of this book to find articles or information about monthly sales planning. In most apparel stores, there are certain given facts: (1) a large percentage of annual

TABLE 5.5 MONTHLY SALES PLAN

MONTH	% OF TOTAL YEAR SALES	SALES PLAN	JUSTIFICATION
April	7.0	$23,100	15% below average due to store opening
May	9.3	$30,690	12% above average due to Mother's Day and new spring wardrobes
June	8.3	$27,390	An average month
July	7.0	$23,100	15% below average due to vacation season
August	8.3	$27,390	An average month
September	9.7	$32,010	17% above average due to back-to-school and career and change of season
October	8.3	$27,390	An average month
November	9.5	$31,350	14% above average due to start of holiday shopping period
December	11.0	$36,300	32% above average; busiest shopping month of the year
January	6.3	$20,790	24% below average due to weather and after-Christmas slump
February	7.0	$23,100	15% below average. Valentine's Day can give a small lift, but bad weather can still be an issue
March	8.3	$27,390	An average month
	100.0	$330,000	

TABLE 5.6 MONTHLY SALES PLAN			
MONTH	% OF TOTAL YEAR SALES	SALES PLAN	JUSTIFICATION
TOTALS	100.0		

sales occurs in the months of November and December when customers are shopping for holiday gifts and clothing; (2) July and January are typically months with slower sales, as customers are usually vacationing in July and broke from the holidays in January. This might be different for your store, however, if your location is in a tourist location (July) or a skiing town (January). As the expert in your product line, you will know how to plan for these highs and lows.

Your profit and loss statement is, in accounting terms, a "fiscal" document, meaning that your annual financial planning will run from the first month that your store opens for a 12-month period, ending in the next year with the month before your store opened. This differs from a "calendar" document, which runs from January to December of one year only. You must decide which month your store will open; this will vary by store type, location, and merchandise offerings. The example below shows our store opening in the month of April; therefore, our fiscal year will run from April of the first year to March of the next year. When you complete your monthly sales plan chart, start with the month that you plan to open your store. Make note, also, that we planned our first-month sales below average because many of our target customers probably will take a while to find and get to know our store. The best approach to planning sales by month is to start with your average months, fill those in, and then work up and down from there. Indicate dollar plan by month, percentage that those monthly sales represent of the total year's plan, and a justification for the planning of each monthly sales (similar to those in the example).

Note: Round off monthly sales plan numbers. Do not use cents; instead, have your sales projections end in 0.

Fixed-Costs Forecasts

At the end of Step 3, you were asked to complete a table of your fixed costs and justify those for your store. Refer to that list and make sure that it is accurate. Perhaps since you prepared that list, you have discovered some corrections that need to be made. Make those now. You will use that finalized corrected list for two forms that we will now prepare: (1) the breakeven analysis and (2) the profit and loss plan. Remember, as you prepare these documents, you will not make a profit if your fixed-cost percent of total sales *exceeds* your gross margin planned percent. The reason for this cautionary statement is that you must pay your fixed expenses from your gross margin dollars.

Breakeven Analysis

Now that you have tabulated and totaled your fixed costs, planned your first-year sales projections, and completed your Gross Margin Exercise, the next step is to complete a breakeven analysis. Before you begin your analysis, you need to put those documents and numbers in front of you.

Bankers, lenders, and the SBA usually want to see a breakeven analysis for your business. This is the chart that determines what gross sales you will need to achieve in order to break even the first year, considering the fixed costs that you will have to pay whether your store makes a profit or not. From that analysis, you and the banker will then be able to estimate the sales over breakeven (will you make money the first year?) that you anticipate, and what profit your store will likely generate in the first year of business. You will want this analysis to help you decide whether or not you need to make some major adjustments before continuing your business plan.

There are three necessary components that you will need to factor into your breakeven equation: (1) planned sales revenue for year one, (2) planned gross margin percent for the first year, and (3) total annual cost of all fixed expenses for the first year.

The basic formula for calculating breakeven is as follows:

Annual fixed costs ÷ Planned gross margin percent = Estimated breakeven sales revenue

$$\frac{\textbf{Annual fixed costs}}{\textbf{Gross margin \%}} = \textbf{Breakeven sales revenue}$$

Next, take your annual planned sales revenue and subtract the breakeven sales revenue from that number to determine your sales over breakeven. Multiply your planned gross profit percentage times that dollar figure and you will get an idea of what profit dollars you may be able to generate in year one.

Table 5.7 demonstrates the breakeven analysis that we completed for our simulated store. Use Table 5.8, which is also featured on the CD-ROM to calculate the analysis for your store.

TABLE 5.7 BREAKEVEN ANALYSIS (completed)	
Annual sales	$330,000
Annual fixed costs	$123,860
Gross profit %	.487
Breakeven sales ($123,860 divided by .487)	$254,333
Sales over breakeven ($330,000 minus $254,333)	$75,667
Profit dollars ($75,667 x .487)	$36,850
Note: Fixed costs and ending profit dollars were rounded to eliminate cents.	

TABLE 5.8 BREAKEVEN ANALYSIS (blank)	
Annual sales	
Annual fixed costs	
Gross profit %	
Breakeven sales	
Sales over breakeven	
Profit dollars	

Profit and Loss Plan

With all the background research completed and the necessary preliminary forms prepared and in hand, you are ready to complete your P & L statements. Many lending institutions require three years of this document: the first year in detail by month, and the second- and third-year plans by quarter. For the purposes of the simulation, we will show you year one by month, and the student project completed plan in the Appendix is also done by month for the first three years of the business. (See Table 5.9 for an example of a completed P & L statement.)

The first step to completing Table 5.10 on page 112 (also on the CD-ROM) is to transfer the fiscal plan months from Table 5.6, "Monthly Sales Plan," onto the first line of the P & L and to enter the sales plan for each of those months just below the name of the month on the line marked "Net Sales." In our example, you can see that we plan to open our store in the month of April, so April is the first month of our fiscal-year plan and the first month listed above the Net Sales line. Your store may be on a different fiscal-year plan, depending on the nature of your merchandise and the key months for your product categories. Enter the months as you have planned them in Table 5.6.

Completing the Top Lines of the P & L

Next, skip down to the Gross Profit line, and enter your overall store gross profit percent that you calculated in Table 5.4, "Sales and Gross Margin Plan." Next, use that percentage to calculate each month's gross profit dollar plan, and enter under the appropriate month on the gross profit line. Total each month's gross profit dollars and get a grand total for the year.

> *Note:* Your annual gross profit (gross margin) dollars on this chart may vary slightly from those on your Sales and Gross Margin Exercise, since you are using rounded-off percentages to calculate monthly gross profit dollars on the P & L. Since everything that you are doing at this stage represents a best estimate, it is not crucial that the numbers match exactly, but they must be very close.

To determine the numbers for the Cost of Goods line, subtract your gross profit percent plan from 100 percent, and you will have your percentage plan for cost of goods. Enter that percentage next to Cost of Goods Sold. Use that percentage to calculate what portion of each month's net sales will be allotted to the cost of purchasing your goods (inventory) and enter that dollar figure under the appropriate month. Remember that cost of goods includes the cost of shipping your inventory from the vendor's warehouse to your store, so you do not need to include those freight costs anywhere else in your plan.

Total all three lines individually (Net Sales, Cost of Goods Sold, and Gross Profit), and enter those figures under the Total column on the far right. Enter percentages as well. If using an Excel spreadsheet, these numbers will appear automatically if you program the proper formulas.

The Middle Lines of the P & L Are for Fixed Costs

Note that we have skipped a line after this and the heading "Itemized Expenses" appears with many entries below. These are the lines for you to enter your fixed expenses. Go back to Table 3.3 (Simulation, end of Step 3) and enter all of your fixed expenses as you have listed them on your Fixed Expenses chart, combining any that are planned to be under $100 per month with another entry. Table 3.3 shows your fixed expenses on an annual basis. Here you will need to list them monthly. Most expenses will show the same amounts each month, but note that budgeted expenses like staffing, travel, and accounting/bookkeeping have different budgeted amounts in some months to account for peaks for these entries.

Determine the Monthly Planned Profit and Loss

At the bottom of the columns for each month, you will need to total all expenses for that month and then subtract those expenses from the gross profit dollars for each month to determine whether you will have a profit or a loss (shown in parentheses). From this exercise, you can see how much of your monthly revenue is spent on inventory and how much is left over (gross profit dollars) to pay your expenses and realize a net profit for the month. You can begin to understand the challenges that a new store owner must face in order to manage a profitable business. When we complete additional financial forms

TABLE 5.9 PROFIT AND LOSS STATEMENT (completed)						
	APRIL	**MAY**	**JUNE**	**JULY**	**AUGUST**	**SEPTEMBER**
NET SALES	23,100	30,690	27,390	23,100	27,390	32,010
Cost of Goods Sold: 51.3	11,850	15,744	14,051	11,850	14,051	16,421
Gross Profit — 48.7%	11,250	14,946	13,339	11,250	13,339	15,589
ITEMIZED EXPENSES						
Staffing	2,100	2,450	2,200	1,700	2,200	2,550
Payroll Taxes	294	343	308	238	308	357
Rent	4,250	4,250	4,250	4,250	4,250	4,250
Electric	330	330	330	330	330	330
Heating, Water	105	105	105	105	105	105
Maintenance	100	100	100	100	100	100
Telephone/Fax	175	175	175	175	175	175
Security/Internet	100	100	100	100	100	100
POS System Maintenance	100	100	100	100	100	100
Insurance	260	260	260	260	260	260
Supplies and Postage	200	200	200	200	200	200
Marketing	1,200	1,200	1,000	800	900	1,200
Travel	100	100	100	100	2,000	100
Accounting	600	300	300	300	300	300
Banking Services	200	200	200	200	200	200
Miscellaneous	200	200	200	200	200	200
TOTAL EXPENSES	**10,314**	**10,413**	**9,928**	**9,158**	**11,728**	**10,527**
NET PROFIT/LOSS	**936**	**4,533**	**3,411**	**2,092**	**1,611**	**5,062**

	OCTOBER	NOVEMBER	DECEMBER	JANUARY	FEBRUARY	MARCH	TOTAL	PERCENTAGE
	27,390	31,350	36,300	20,790	23,100	27,390	330,000	
	14,051	16,083	18,622	10,665	11,850	14,051	169,289	51%
	13,339	15,267	17,678	10,125	11,250	13,339	160,711	49%
	2,200	2,500	2,900	1,500	1,700	2,200	26,200	8%
	308	350	406	210	238	308	3,668	1%
	4,250	4,250	4,250	4,250	4,250	4,250	51,000	15%
	330	330	330	330	330	330	3,960	1%
	105	105	105	105	105	105	1,260	0%
	100	100	100	100	100	100	1,200	0%
	175	175	175	175	175	175	2,100	1%
	100	100	100	100	100	100	1,200	0%
	100	100	100	100	100	100	1,200	0%
	260	260	0	0	0	0	2,080	1%
	200	200	200	200	200	200	2,400	1%
	1,100	1,300	1,500	900	1,000	1,100	13,200	4%
	100	100	100	100	2,000	100	5,000	2%
	300	300	300	300	300	600	4,200	1%
	200	200	200	200	200	200	2,400	1%
	200	200	200	200	200	200	2,400	1%
	10,028	**10,570**	**10,966**	**8,770**	**10,998**	**10,068**	**123,468**	**37%**
	3,311	**4,697**	**6,712**	**1,355**	**252**	**3,271**	**37,243**	**11%**

TABLE 5.10 PROFIT AND LOSS STATEMENT (blank)

MONTH	APRIL	MAY	JUNE	JULY	AUGUST	SEPTEMBER	
NET SALES							
Cost of Goods Sold: 51.3							
Gross Profit — 48.7%							
ITEMIZED EXPENSES							
Staffing							
Payroll Taxes							
Rent							
Electric							
Heating, Water							
Maintenance							
Telephone/Fax							
Security/Internet							
POS System Maintenance							
Insurance							
Supplies and Postage							
Marketing							
Travel							
Accounting							
Banking Services							
Miscellaneous							
TOTAL EXPENSES							
NET PROFIT/LOSS							

	OCTOBER	NOVEMBER	DECEMBER	JANUARY	FEBRUARY	MARCH	TOTAL	PERCENTAGE

in later steps, such as the cash flow statement and income statement, you will see the effect of additional deductions from your profit line like taxes, repayment of loans, and owner salaries.

Profit and Loss Plans for Years 2 and 3 for the Business

You will also need to complete P & L statements for year 2 and year 3 for your business. For students, it is recommended that you complete year 1, perhaps have the professor check it, prepare your other financial documents for year 1, and then go back and prepare years 2 and 3 for all financial documents. Since you have not yet been required to complete your marketing plan and your personnel plan, you may find that you need or want to make changes to your P & L after you have further analyzed these key expenditure categories. Waiting to complete subsequent years of the P & L will allow you to make changes if necessary to year 1 without having to make revisions to the other years as well. For entrepreneurs, we would recommend the same procedure, but you might want to have your accountant check your work before proceeding to complete the documents for all three years.

When completing the P & L for years 2 and 3, we would suggest that you keep your sales plan increase at a conservative level until your store opens and you are able to ascertain the rate of customer acceptance of your store and product. Keep in mind that your business plan, and particularly your P & L and your cash flow statements, should be frequently reviewed and updated even after your store is operational. If you open to booming success and you see that level of customer acceptance continue for a few months, you can go back and revise those documents to allow for faster growth.

After analyzing your growth potential for years 2 and 3, decide upon a percentage of increase for your net sales in those years, and rework your monthly sales plan accordingly. Make note of fixed expenses that will probably increase in those years. Your lease will state whether or not you will experience an increase in monthly rental costs, but this will likely not happen in the first three years. Salary costs and taxes on salary will probably need to increase as your sales plan increases. The utility companies can give you an idea about whether to expect those costs to rise, as can the telephone company. You can allow for unexpected increases by upping the budget for miscellaneous expenses.

Simulation

1. Using the information in Table 5.1 as a guide, formulate your gross margin assumptions for your store. Refer back to the work that you did at the end of Step 3 for your first-year annual sales plan. Enter the assumptions into Table 5.2
2. Complete the Sales and Gross Margin Exercise in Table 5.4.
3. Determine your monthly sales plan and put that into Table 5.6.
4. Refer back to Step 3, Table 3.3, for the fixed costs that you listed there. Fine-tune the list and extend it from the annual costs that you figured initially into monthly expenses for each fixed cost.
5. Prepare a breakeven analysis and put the information into Table 5.8.
6. Use Table 5.10 to complete a first-year P & L statement, such as the one in Table 5.9.
7. Write justifications for your fixed expenses and P & L statement to explain costs and how you plan to minimize/control them. You don't need to justify fixed expenses twice; merely update any justifications that you started in Step 3.
8. Refer to the sample business plan in the Appendix, Financial section, for an example of the way that fixed expenses, Gross Margin Exercises, breakeven analysis, and P & L plans can be completed.
9. File these completed forms under "Financial Documents" to be included in that section of your final business plan.

STEP 6
Merchandising Your Store

"The difference between style and fashion is quality."

—Giorgio Armani

PROJECT OBJECTIVES

This step in the simulation will guide you through:

- Selecting the right product for your store and target customer
- Maximizing inventory levels through smart assortment planning
- Understanding the keys to managing your in-store assortment in order to maximize both profit and customer satisfaction
- Planning and calculating an opening stock assortment for your new store
- Using buying trips effectively for new stock purchases and gathering information about the industry and market

Some might argue that, while financing and marketing are key parts to the success of any new retail venture, without the right product at the right price in the right place aimed at the right target customer, there can be no business. Certainly it is true that the smart owner will become an expert in both product assortment and what motivates the target customer to make a purchase. According to Deidra Arrington, past vice president of Peebles Department Stores, "Inventory is a retailer's biggest asset, yet many business owners find the prospect of planning inventory a daunting task. . . . [However,] it is critical to improving sales, profit, and turnover." By the time you reach the end of this step, you will feel more comfortable about the whole process.

Planning by Department and Classification

In Step 1 you selected the categories of merchandise that you will carry in your new store. In retail planning, merchandise is divided into ***departments*** and ***classifications***. In a big department store, for example, you may find many departments: men's, women's, children's, home, and so forth. Each of these departments could technically stand alone as a specialty store. Within each department, merchandise is divided into product classifications (sometimes called categories). Table 1.1 in Step 1 lists some examples of typical classifications that can be found in specialty stores. Keep in mind that while large department stores with tens of thousands or even hundreds of thousands of square feet can afford to offer many departments and classifications, a small store with limited space cannot. The smart owner selects one or two departments and limits classifications so as to offer some depth of product in each of those classifications. A store that tries to offer only one or two pieces of many different items and classifications will only confuse and frustrate its customers.

After determining merchandise classifications or product categories, create your own system for keeping track of ***purchases***, ***receipts***, sales, markdowns, and vendor returns. One way to accomplish this is to assign codes for each type of merchandise that you offer in your assortment. These can be broken down further than your product classifications but should not be tracked by individual item. In a small, trendy boutique, you should not be concerned about tracking sales, markdowns, and so on for each item; that will create too much paperwork and will not be an asset to the business. Within the accessory classification, for example, you might want to assign a code and keep track of sales for jewelry versus handbags, but you would not want to keep records on how the messenger bag in black sold versus the tote in brown. Handbag styles, like most fashion items, change every season. It is wise to plan total accessory sales by category only; you will then have the flexibility to increase your stock in, for example, hats for next season if they are going to be a trend and decrease your inventory in sunglasses if that class of goods is trending down.

Buying for a New Store

A new store owner might have legitimate concerns about how to select merchandise for a store that is opening its doors for the first time. After the store has been open for six months or a year, the owner will have a much clearer understanding of what appeals (and what does not) to the consumer and can use that information to make future purchases. However, the initial inventory selections should be made based upon the following information:

- The owner/buyer must put aside personal preferences when selecting merchandise to sell in his or her store. Unless you and your friends fit the target customer profile exactly, don't worry about what you would or would not wear when selecting product. Do your research to learn what your target customer will buy and what motivates him or her to make a purchase. Now that you have developed a profile for your target customer, always keep that profile in mind when making your selections. You cannot be all things to all consumers, and you cannot attract opposing consumer types to your small store—for instance, the rich society lady will not shop where the assortment is keyed to the working-class middle-income mom, even if you do try to sneak in some high-priced items.
- In addition to the trend research that you will naturally conduct, gather as much information as possible from those who are experienced in the type of products that you will carry. Solicit input on what sells and what does not in stores that target a similar customer. Ask for advice and observe firsthand what customers purchase in similar stores.

- Ask your vendors. They generally have your best interests in mind. After all, a good ***sales representative*** (often called a "rep") will want you to be successful with his or her products so that you will come back to purchase again. While you certainly don't want to leave all decisions up to the reps, listen and evaluate the important information that they have and benefit from their firsthand experiences. They will probably even give you information about what sells and what doesn't in the lines of their competitors.
- Before you go to market or to attend a trade show, make a financial assortment plan for your opening inventory. This will help you stay focused and avoid overbuying in one category while forgetting another. At the end of this step, you are asked to create such a plan. You can find an example of this in the Appendix, Part V Product and Service of the sample business plan.
- When you go to market or shop the assortments of local sales reps, don't buy the first items that you see—even if you really like them. Take the time to shop around and see what is available before making the critical decisions about the items in your opening inventory (Figure 6.1). Comparing before buying also allows you to better plan your inventory—to make sure that you have all important items covered and that you haven't purchased too much duplication of items.
- Identify ***key items*** in each merchandise category, items that you determine to be the ones that will bring volume sales and have low risk. Make sure to buy more units on these commodity items, and then attempt to stay in business at all times with

Figure 6.1
It is important that a buyer makes informed purchasing decisions.

key colors and sizes in those items (or the next incarnation of those items). These can be the basis around which you plan all inventory in that category. For a lingerie store, for example, these items might be a cotton panty and a best-selling underwire bra (Figure 6.2). Your key items, however, must always be aimed at your target customer.

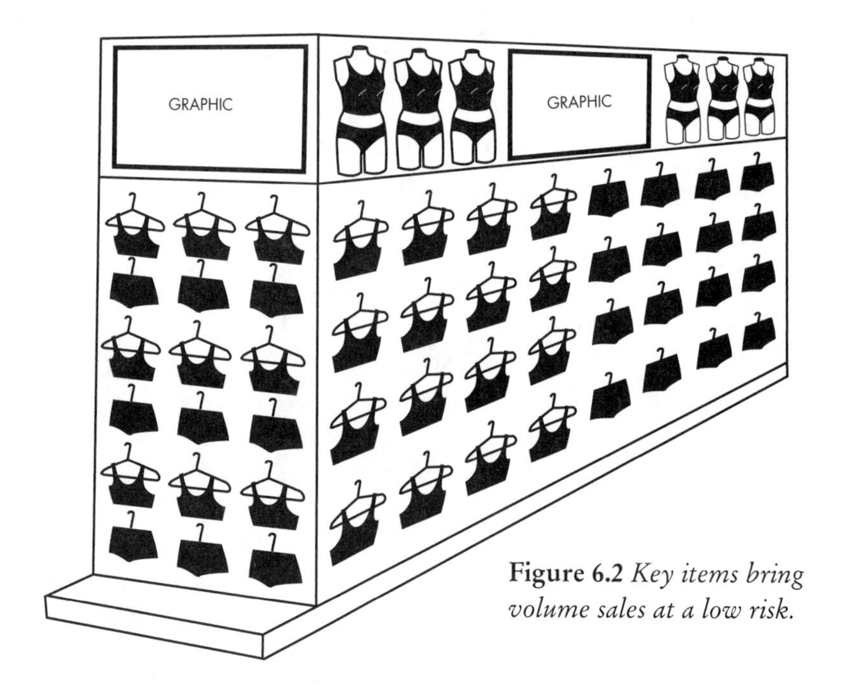

Figure 6.2 *Key items bring volume sales at a low risk.*

- If you find items that you feel are going to be important future trends but you are unsure of your customer's readiness for them, buy small quantities to test their appeal.
- Buy with a thought to ***visual merchandising***: Where will you display these items or branded lines in your store? Will the colors complement other colors that you plan to carry? You don't want to purchase tops in colors or styles that will not complement the bottoms that you plan to offer. Your aim should be to sell outfits to your customers; entice her or him to buy multiple items.

- Do not allow enthusiasm and excitement for the merchandise to cause you to overbuy. This is one of the biggest mistakes that buyers and owners make. A key truth in buying wisdom is this: *It is always better to be underbought than overbought*. This will allow for the unknown. Suppose, for instance, your sales aren't as strong as you anticipated. Or what if a new trend or a new fad emerges and you need reserve funds to purchase it? Or suppose you decide to try merchandise from a local artisan. Excessive merchandise almost always results in excessive markdowns and lower profits.

Assortment Planning

Deidra Arrington calls the six-month merchandise buying plan "a roadmap for success." This buyer's tool uses sales, markdowns, and purchases to create a comprehensive guide to planning, selling, and buying store stock that will help the buyer avoid those awful traps that can and will erode margin. The traps are overbuys that result in ***overstocks*** and slow turnover that lead to markdowns, and the opposite "cousins," underbuys, which lead to understocks and too rapid a turnover, resulting in unhappy customers and lower sales for the store. Both circumstances end in reduced profits. Proper merchandise planning is crucial for all retail stores both large and small, but it is especially critical for the small store owner who has less margin of error between success and failure. It is always surprising to us when we speak to small store owners and realize how many of them do not use six-month planning tools to plot their inventory. In fact, some stores seem to have no plan at all except to go to market and buy what they think they can afford and what looks good to them at the time. Of course, no amount of planning will ensure success unless the plan is followed, monitored, and updated frequently as the market and store circumstances change.

Like any worthwhile venture, the process of creating and managing the six-month buying plan is time-consuming, but it is well worth the effort. The plan can work at any level: by total store, classification, department (for a larger store), or vendor. Some stores (shoe stores, for example) may find certain vendor products are so important to their total assortments that they may want to plan those selections by vendor instead of category.

Everything in the plan originates with good monthly sales projections; you have already done that for your store (Step 5). After your store has been open for a couple of months and you have some good solid data on sales, inventory, and slow sellers, you should employ the six-month buying plan as your primary tool in ***inventory control.*** You will be able to look back on sales results from your first months in business, noting what sold well and what had to be marked down. You can then take the necessary steps to either quickly find more merchandise; ask that merchandise that has been purchased for later delivery be shipped to you sooner; or if things don't go as planned, ask if merchandise can be delayed in shipping due to a slower-than-anticipated opening. After a year in business, this planning process will become even more valuable as you will think back to periods when you did not have enough stock to make your sales plan for the month and satisfy customer needs, and you will also remember those dark days when you had so much inventory that the fixtures were bulging and you had no recourse but to slash prices. With this information, you must also keep in mind that it is impossible to buy everything that customers ask for, but you will have had time to figure out which of those missing items would add significant incremental sales to your store and which ones would gather dust. By the second year, you would have taken a physical inventory to see which categories of merchandise are left over and which are completely out of stock; this will assist you in recognizing and acting on missed opportunities.

One of the most important advantages to six-month planning is the ability to readily see by month how your inventory investment for that month translates into sales. Even if your inventory for the year is on target with your plan, are you flowing the receipts of that merchandise by month according to the way that your sales are happening?

Some of the tools used in the planning and execution of the six-month

PERRY'S

SIX-MONTH DOLLAR PLAN

DEPT. NAME ______________________ DEPT. # __________

BUYER ______________________

FALL		AUGUST	SEPTEMBER	OCTOBER	NOVEMBER	DECEMBER	JANUARY	FEBRUARY	SEASON TOTAL
SALES $	Last Year	$ 225.0	300.0	210.0	255.0	390.0	120.0		1,500.0
	Plan								
	% Inc/Dec								
	Revised								
	Actual								
STOCK/SALES RATIO	Last Year	3.9	3.0	4.1	3.6	2.5	6.3	5.7	
	Plan								
BOM STOCK $ (Retail)	Last Year	$ 877.5	900.0	861.0	918.0	975.0	756.0	850.0	876.8 avg.
	Plan								
	Revised								
	Actual								
MARKDOWNS $	Last Year	$ 89.5	45.0	66.0	51.5	134.0	64.0		450.0
	Plan								
	% to Sales								
	% by Month								
	Revised								
	Actual								
PURCHASES $ (Retail)	Last Year	$ 337.0	306.0	333.0	363.5	305.0	278.0		1,922.5
	Plan								
	Revised								
	Actual								

SEASON TOTAL	LAST YEAR	PLAN	ACTUAL
Sales	$1,500.0		
Markup %	55%		
Markdown %	30%		
Gross Margin %	41.5%		
Average Stock	876.8		
Turnover	1.71		

NOTES:

Figure 6.3 *A six-month buying plan.*

buying plan are monthly sales plan, stock-to-sales ratio, inventory turnover, planned EOM (end-of-month) stock (inventory), planned markdowns for each month, and planned BOM (beginning-of-month) inventory. With this data, you will be able to set up your charts and calculate purchases needed for each month. Another name for purchases is receipts, and it is important not just to buy merchandise but to know when you should receive it into your stock (Arrington, *Planning Inventory the Right Way*). See Figure 6.3 for an example of a six-month buying plan. For detailed information on how to calculate and use the six-month buying plan, we recommend an excellent text: *Perry's Department Store: A Buying Simulation* by Karen M. Guthrie and Cynthia W. Pierce (Fairchild Books, Inc.).

Creating an Opening Assortment Plan

The six-month buying plan is a financial plan; the buyer/owner must still complete an assortment plan by category or style that will serve to show which merchandise will be received into stock in which month in order to maximize sales. For this simulation and for the completion of the business plan for a new store start-up, it is only necessary to complete an assortment plan by category. The financial community who will help to finance the business is not interested in every single sweater or skirt that you plan to purchase for your opening. They merely want to see that you have a plan by category for the merchandise that will be in your beginning inventory and that it adds up to a dollar purchase figure that makes sense for the business. Once you have completed this plan by category, you will take it with you on your buying trips so that you will have a plan for how many units to buy by category and price point. Of course, all plans must be somewhat flexible and you will finalize your opening inventory purchases based on the best merchandise that you find. Later in this step, we will discuss how to complete an opening inventory chart by category, and you will be directed in the simulation steps to prepare a chart for your own store.

Keys to Planning and Managing a Successful Business

There are four key operating ratios that will aid you in planning and controlling your business. We have found that many small store owners do not understand or use these ratios, but you will be ahead of the game for your business if you begin to incorporate these tools into the planning and monitoring of your inventory.

Sales per Square Foot

We discussed and used this ratio in Step 3 when we projected the first-year annual sales estimate for your business. This is a very useful tool that most stores, large and small, use to evaluate the success of their businesses. The premise behind the formula is that you must understand what the space in your store costs to operate each day. After all, you will pay many fixed expenses for the use of that space: rent, lighting, heating, insurance, and on and on. It is therefore crucial that you use that space to best advantage and realize the most desirable sales revenue for each square foot of space inside your store. Certainly, it is true that some stores will have a great deal of space between fixtures and in store aisles, while others will cram merchandise into nearly every square foot of space inside the store—to the extent that mobility between fixtures becomes a real problem. Part of the decision about space between fixtures and width of aisles is related to the price points of your merchandise and the ambience that you feel that you must project in order to attract and hold your target customer; we will discuss this further in Step 9 when we cover visual merchandising. The hard truth, however, is that you cannot afford to carelessly waste space within your store. Inexperienced retailers will often fail to plan space usage properly and therefore end up with either too much merchandise that results in a cluttered look or too little that keeps them from realizing their full sales potential.

Selling Square Footage

Your total square footage for your store is reduced by the footage of non-selling space to give you a more usable figure of "selling square footage." To get to the selling square feet of your store space, subtract the square footage for parts of your store that do not contribute directly to your sales effort. These include the stockroom, office space, the area behind the cash register, and sometimes the dressing rooms. Typically, dressing rooms are counted as selling space if they have mirrors (supposedly to aid in the selling effort) and are deducted from selling space if they do not (Dion and Topping, p. 49). This can work in reverse, however, and the decision whether or not to have mirrors in your dressing rooms can be part of a selling strategy. Chicos, for example, has no mirrors inside dressing rooms; as customers, we have concluded that this decision was made so that women will be forced to come out of the dressing rooms to look in large mirrors where a salesperson is there to suggest belts, jewelry, tank tops, and other add-on purchases. Regardless of this deviation from the norm, the basic formula for calculating selling square footage remains.

As a refresher, the formula for calculating sales per square foot is as follows:

$$\frac{\textbf{Net annual sales (or sales plan)}}{\textbf{Total selling area in square feet}} = \textbf{Sales per square foot}$$

In Step 3, we showed the example of how we calculated sales per square foot for our simulated store:

$$\frac{\textbf{\$330,000 (first-year sales plan)}}{\textbf{2,200 square feet of selling space}} = \textbf{\$150 per square foot plan}$$

Variance to the Plan

Sales-per-square-foot numbers will vary greatly among different product lines and certainly from store to store. Factors such as how long a store has been in business, what type of merchandise it carries, and even the size of the store will impact the results. Large stores such as department stores often use this ratio as a way to judge results from department to department and to help determine whether or not space within the store should be shifted and altered to allow more room for strong-performing lines of merchandise. As a small store owner, you will want to do much the same: monitor your annual results to ascertain whether or not you need to adjust inventory investments.

You can research actual sales-per-square-foot results in specific stores at BizStats.com to help you determine what your goals should be, but look at those results with the understanding that most of these stores have been in business for many years and have had an opportunity to establish a customer following. You should plan on the conservative side until the consumer finds and develops loyalty to your operation.

Turnover

Another very important tool that we will use in the simulation is a ratio that will show us the rate at which our average stock is selling in a given time period, allowing us to reinvest the money from those sales into new inventory, or "how many times inventory is 'turned' into cash. Turnover is an indicator of the efficiency of inventory" (Arrington, *Planning Inventory the Right Way*). The formula for calculating turnover is as follows:

$$\frac{\text{Net sales (or planned sales)}}{\text{Average inventory}} = \text{Turnover}$$

Conversely, if we know the desired turnover that we want to achieve for our new store, we can determine the value of the opening inventory in this manner:

$$\frac{\text{Net sales (or planned sales)}}{\text{Planned turnover rate}} = \text{Opening inventory value}$$

The two equations above would give us, respectively, a turnover rate and the value of inventory expressed as *retail* dollars. Turnover can be calculated at both retail dollars and cost dollars (COGS, or the amount of money that you will spend to buy your inventory). For the next part of the simulation, when you complete the Opening Inventory chart, you will need to be most concerned about the calculation of inventory dollars at cost, and you will use a planned turnover rate in order to do this.

You might ask what turnover rate should a new store plan? Dun & Bradstreet Credit Services (www.dnb.com) is one place you can go to research typical results for different kinds of stores. You might also find articles about small boutiques in publications like *Women's Wear Daily* or your local newspaper, that refer to store turnover rate and also sales-per-square-foot data. Your turnover plan should match your product line, but for most small apparel stores, you should plan your first-year turnover rate at between 3.0 and 4.0. Turnover is expressed in tenths of a percent, and a buyer/owner is rated as doing a good job if he or she is able to increase turnover by even a tenth of a percent (for example, from 3.4 turns to 3.5 turns).

Turnover as an indicator of success can also be a tricky ratio. If your turnover results are too low (below 3.0 for most apparel stores), then you are probably holding on to out-of-date merchandise for too long, or merchandise that the customer has voted that he or she does not want. A slow turnover is an indication that your inventory is stale and that you aren't freeing up inventory dollars fast enough to allow you to reinvest that money into newer, more exciting products. A turnover that is too high might indicate that you are frequently out of stock on key sizes in key items. This is especially true for classic or basic items. In a lingerie store, if we were to calculate turnover for the bras, it would show a slower turnover than for the trendy sleepwear items. The reason is that when you are in the bra business, you must carry more inventory (***stock-keeping units***, or ***SKUs***) for different styles on a regular basis to accommodate customer demand. You would also want to keep those key sizes and colors in stock on your best-selling items, so you would need to reorder that merchandise more frequently. Trendy sleepwear items, on the other hand, will change from season to season, and it is not always necessary to carry all sizes and colors. You would also mark down or reduce the price of seasonal items at season end and slow sellers within the season to allow for the purchase of newer, more desirable styles. Thus, your trendy sleepwear category would turn more quickly than your more basic bras. No retailer can afford to be in stock 100 percent of the time on all categories, so you will need to determine which categories require a higher amount of inventory on hand (more basic items) and which ones you are willing to let run out in a shorter amount of time. In this way, you will be able to manage the optimum turnover rate for your store.

To proceed to the calculation of opening inventory for our simulated women's apparel store, we checked the Dun & Bradstreet results for Apparel, women's ready-to-wear, and found the median rate of turnover to be 4.4. A look at Apparel, women's specialty stores, showed the median rate of inventory turnover to be 3.8. Because we are a new store and must woo and win our customers, we decided to take a conservative approach and plan our turnover rate at 3.5. You should now do some research and make some decisions about your turnover plan. You will need this figure to complete the simulation instructions at the end of this step.

Stock-to-Sales Ratio

This indicator is related to and works with turnover. It helps the owner determine whether or not the inventory level is in the right amount to maximize sales and minimize markdowns. This ratio is expressed as a fraction; for example, if the stock-to-sales ratio is 4/1 (four to one), then the store is turning inventory at a 3.0 rate or three times a year. This would mean that each month the owner would have in stock four times the amount of inventory that he needs to generate planned sales (Dion and Topping, p. 46). These are, of course, financial ratios and do not mean that the store ever sells completely out of merchandise. That would happen only if the owner was ready to close the doors.

Gross Margin Return on Investment

Gross margin return on investment (GMROI, also sometimes referred to as gross margin return on inventory investment, or GMROII) will give the owners and financial backers of a business an idea of how much return they are realizing on the business. Similar to the way a financial planner might track the return on a stock or bond fund, GMROI indicates how much they are getting back for every dollar invested in inventory (Dion and Topping, p. 47).

GMROI is always expressed as dollars; for example, a GMROI of $3 would mean that you are receiving three dollars back in sales for every dollar invested in inventory.

For this simulation, we will only use the first and second ratios discussed, sales per square foot and turnover. As a store owner, however, you should be familiar with the other two ratios. You will use the stock-to-sales ratio when preparing your six-month buying plan, and your GMROI will be a good indicator of whether or not you need to shift receipt of goods to accommodate the highs and lows of sales activity.

Completing an Opening Assortment Chart

To complete this chart, you will need to refer back to the Sales and Gross Margin Exercise that you completed in Step 5 (Table 5.4). You will also need a copy of your Profit and Loss Statement that you completed for year one (Table 5.10). Make note of the annual turnover plan that you determined to be the best estimate for your store. You will want to have in front of you the list of key price points for each of your product categories with weighted average retails. (You completed this exercise in Step 5 as well.)

As you can see in Table 6.1, the product categories that you used for your sales and gross margin analysis should be listed down the left side. Across the top, you should include a column for average retail price points for each of your product categories, and one to show average cost prices for each category. This information can be extracted from your Sales and Gross Margin Exercise. You will not need to calculate cost prices for each of your price points, as this chart is merely a starting point for the purchase of your inventory and you will fill in actual numbers after you make your purchases. Next to that is a column to show the percentage of each category's business that each of your price points will represent. For example, in the chart for our simulated store in the category for dresses and skirts, the most important price point will be $125, and that will represent 40 percent of the business in that particular category. The percents for each price point in each category should add up to 100 percent before you move on to the next category. (You calculated this part of the exercise in Step 5.)

Before completing the rest of the chart, you need to determine the cost value of your opening inventory. Go back to your P & L statement for year one and look at the total annual dollar plan for cost of goods sold (COGS). This figure represents the total annual investment that you will need to make when purchasing the goods to sell in your store. Remembering the explanation earlier in this step about turnover and what it designates, take your total annual plan for year one COGS and divide that by planned turnover. For our simulated store, our formula looks like this:

$$\frac{\textbf{\$169,289}}{\textbf{3.5 planned turnover}} = \textbf{\$48,368 value of opening (average) inventory}$$

As with all our planned numbers, we want to round this off to $48,000 at cost. The reason for our fixation on the value of inventory at cost is that this

TABLE 6.1 OPENING INVENTORY PLAN (completed)									
CATEGORY	**AVG. RETAIL**	**AVG. COST**	**% PRICE PT.**	**#PC./PRICE PT.**	**PLANNED %**	**TOTAL COST**	**ACTUAL PC.**	**ACT. COST**	**ACT. RETAIL**
Dresses/	$100.00		35%	75					
skirts	$125.00		40%	86					
	$160.00		15%	32					
	$200.00		10%	22					
Wgt. avg.	$129	$55.76	100%						
TOTAL				**215**	**25%**	**$12,000.00**			
Bottoms	$70.00		20%	39					
	$100.00		25%	48					
	$125.00		35%	68					
	$150.00		20%	38					
Wgt. avg.	$113.00	$49.85	100%						
TOTAL				**193**	**20%**	**$9,600.00**			
Tops	$45.00		25%	106					
	$60.00		40%	170					
	$80.00		25%	106					
	$100.00		10%	43					
Wgt. avg.	$65.00	$28.25	100%						
TOTAL			**30%**	**425**	**25%**	**$12,000.00**			
Jackets	$125.00		30%	31		38			
	$160.00		40%	42					
	$200.00		20%	21					
	$250.00		10%	10					
Wgt. avg.	$167.00	$69.14	100%						
TOTAL				104	**15%**	**$7,200.00**			
Access.	$30.00		35%	127					
	$50.00		40%	146					
	$75.00		15%	55					
	$100.00		10%	36					
Wgt. avg.	$52.00	$19.80	100%						
TOTAL				**364**	**15%**	**$7,200.00**			
GRAND TOTAL				**1301**	**100%**	**$48,000.00**			

is the amount of money that we will have to budget to purchase opening inventory (after all, the store owner pays for inventory at cost and then sells it at retail dollars) when we make our list of capital expenditures (start-up costs) in Step 10.

With this information, you should be able to work backward to determine the number of units that you plan to purchase by price point by category, and to get total cost and total retail value for each category. In the first column, you will see a designator for average retail price for each of your categories. We have rounded these off on the Gross Margin Exercise. Note that the weighted-average retail for dresses and skirts calculated at exactly $129, but the weighted-average retail for bottoms came to $112.75. Since that is not a retail price point that would be used in the industry and since this exercise is an estimate for planning purposes only, we rounded that number off to $113.

From the Sales and Gross Margin Exercise (Table 5.4), copy the average cost from line 3 of each category. Then enter these costs in Table 6.2 on page 134 and on the CD-ROM under the column marked "Avg. Cost" and on the corresponding lines marked "Wgt. Avg."

We can determine the total number of units per category by using the percentages that we listed on the Sales and Gross Margin Exercise. We said that dresses and skirts account for 25 percent (regular-price plus sale-price goods) of the inventory, so if we take our opening inventory value at cost ($48,000) and multiply this times the percent value of the category (25 percent), we determine that:

$$\$48{,}000 \times .25 = \$12{,}000$$

This is the value at cost of the opening inventory for the dress/skirt category. We can perform this action for all categories. Using the percentages that we have listed in column #3, we determine, for example, that the $125 retail price point is worth 40 percent of our business, so we will spend .40 × $12,000

TABLE 6.2 OPENING INVENTORY PLAN (blank)									
CATEGORY	AVG. RETAIL	AVG. COST	% PRICE PT.	#PC./PRICE PT.	PLANNED %	TOTAL COST	ACTUAL PC.	ACT. COST	ACT. RETAIL
Wgt. avg.									
TOTAL									
Wgt. avg.									
TOTAL									
Wgt. avg.									
TOTAL									
Wgt. avg.									
TOTAL									
Wgt. avg.									
TOTAL									
GRAND TOTAL									

on the purchase of goods for that price point, resulting in a figure of $4800. At this point, you need to determine the total number of pieces (#Pc./Price Pt. on chart) that you plan to purchase for each category. Divide the total cost (e.g., dresses $12,000) by the average cost for each category (e.g., dresses $55.76) and round off the totals. Our dress category will look like this:

$$\frac{\$12{,}000 \text{ total cost}}{\$55.76 \text{ average cost for category}} = \text{215 pieces (rounded) to purchase in total for the dresses category.}$$

Then use the "%Price Pt." column to determine how many pieces to purchase at each price point. For example: at the $100 retail price point, we would purchase 215 total pieces (for the category) × 35% at the $100 price point, resulting in a plan of 75 units to be purchased at that retail price. Do this exercise for all categories and all price points. Then add all category totals to get the total units to purchase for your store opening. Enter this total on the last line of the form, marked "Grand Total."

We could work backward again to find the retail prices of our goods as they will first be offered to the customer. To do this, we need to use the initial markup percentages that we established on the Sales and Gross Margin Exercise chart. For that same category of goods (dresses and skirts), we determined that our initial markup is 56 percent. Using the reciprocal of the markup, the total retail value of our $12,000 purchase at cost would be $12,000 divided by .44 (44%), or $27,273 (rounded). After you complete all columns in this manner, you will be able to see how many sales dollars you *would* be able to generate if you sold everything at regular price. Alas, that cannot be, as stated previously. The real sales dollars will be significantly less when you factor in markdowns taken. For this exercise, it is not required that you compute retail price dollars, but it is an interesting exercise if you wish to see what your inventory would be worth if you sold everything at full price.

Note that in Table 6.2 at the far right are three columns: Actual # Pieces (purchased), Actual Cost of those units, and Actual Retail (full retail value) of those pieces. Since this is a worksheet for you to use when you place your buys, it is not necessary that you fill in those columns now. You will fill them in later after you place your orders to buy your opening inventory.

The value of this plan is to show how you will plan your opening inventory purchases—that is, where the emphasis will be in terms of categories and price points. You certainly would not want to plan to retail your merchandise at 20 different price points from low to very high. By completing this chart, you can put down on paper in a logical manner where your key retail price points will be and make a plan that you can take with you to market. As you will observe, this plan is done neither by style nor by color or size. It gives the buyer/owner flexibility to make those decisions after the market or trade show visits and after he or she has had the opportunity to evaluate the brands and styles that are available to purchase for the grand opening. It also lets your potential financial backers see that you understand the importance of a plan for your inventory and that you are prepared before you go to market and spend your (and their) money in an organized fashion.

Fashion Research

Before you embark on any visit to a market or trade show to purchase inventory, you must be certain that you are up-to-date on new trends in your product categories. The chances are more than strong that you are very interested in, or have direct experience with, any product categories that you plan to carry in your store. Perhaps you have worked in a store that offers similar merchandise. Perhaps you are a recent graduate of a fashion merchandising program where you studied and observed fashion trends. Perhaps it is just a strong personal interest that has kept you informed and aware of updates in the world of fashion. Regardless of past experience, however, it is crucial that you know something about future trends before you spend your money on inventory for a new store. Fashion trends are relative, of course, to the willingness of the consumer in your local area to accept them. Just because the runway models and the fashion-forward celebrities are wearing something does not necessarily mean that the "ladies who lunch" in Richmond, Virginia, will be ready to make it part of their wardrobes. Here we must relate back to the information that you gathered about your target customer in Step 1. Refer to the list of attributes you created about your intended customer base, and try to put yourself in the mind of your "typical" target customer. For example, what are her daily needs? How many different wardrobes is she likely to need: one for work, one for play, one for formal occasions, one for vacation? Where will her focus be and what are the most important items that she is likely to seek and buy from your store? How does she allocate her disposable income? Does she focus on the kids and think of her own needs last, or is she a society woman who must have a different outfit for every occasion?

Regardless of the final decisions that you make about which items will work in your store and for your customer, it is important to be as up-to-date as possible on the direction of fashion trends. How should you accomplish this? Ideally, you would subscribe to the services of one of the many excellent trend forecasting services such as Promostyl of Paris and New York, or Label Works for the young trends, or Carlin of the West Coast for men's wear. These companies offer the most up-to-date information about future trends, but their products are very expensive and the average new store owner cannot afford to subscribe. The following are some ways that you can stay up-to-date on what are the most important colors and styles for your customer:

- Subscribe to *Women's Wear Daily* if you are in the women's apparel business, *Daily News Record* if your lines are for men, *Children's Digest* if you cater to the very young set, or *Home Fashions News* if your store will offer fashion for interiors (Figure 6.4). These are only some of the excellent trade publications available to those who are involved in the fashion business. They will tell you about trends for next season, the season after that, and beyond, and will keep you aware of the needs and desires of the fickle fashion consumer.

VALENTINO ON SHOW/3 THE OLSENS' NEW JEWELS/12

Women's Wear Daily • The Retailers' Daily N... June 16, 2008 • $2.00

WWDMONDAY

Accessories/Innerwear/Legwear

Fantasy Island

Lanvin's Alber Elbaz glanced back to a Thirties glamour girl while injecting whimsy with fun embroideries and glitzy costume baubles. "It's about the original idea of cruise, a fantasy vacation," Elbaz said. One example: the playful cotton T-shirt and silk skirt combo shown here. For more on resort, see pages 4 to 6.

Coming Under Pressure: Activist Shareholders Target Retailers Again

By Evan Clark

THE GADFLIES ARE BACK — and this time they've got deep pockets.

Target Corp., Dillard's Inc., Chico's FAS Inc., Charming Shoppes Inc., Syms Corp. and Haggar Corp. are just some of the fashion firms that have faced, or are confronting, activist shareholder campaigns. These outsider efforts are often marked by lawsuits, battles to sell divisions or replace board members and the airing of as much dirt as can be excavated.

It is all part of the deal when companies tap the public equity markets and open themselves up to

See **Retailers**, *Page* **15**

Figure 6.4 *Trade publications, such as* Women's Wear Daily, *publish the latest fashion trends.*

- Read everything. You will find good fashion information in many fashion magazines, but you can also read articles about fashion in the local newspaper, any local free press (*Style, Inc.* is a particularly good one in Richmond, Virginia, but you probably have some in your town also), and town or regional magazines; even specialty publications like the *Wall Street Journal* have wonderful articles every week about trends in apparel, accessories, home, and technology. It seems that everyone today is interested in fashion. There is a difference, of course, in the quality of the information among these different sources. Even with fashion magazines, some (such as *Lucky* and *In Style*) are primarily shopping magazines that report what is in the stores now and are aimed primarily at teens and young women. Other fashion magazines report and show pictures of current trends but also give a taste of future trends—or trends that are just catching on in big cities like New York or Los Angeles but might not be ready for cities like Richmond, Virginia. Some examples of these are *Vogue*, *Bazaar*, *Elle*, and *W* (Figure 6.5). With time and study, you will begin to pick up signs of future trends as you study and observe current ones. Most important, you will get a feel for which styles are fads (short-lived fashions that will probably be gone in a season or two), which are trends (longer-lived fashion items that are likely to be around for several years), and which are in the category called classics or basics (apparel that is never really out of style but may just change slightly from season to season—like the basic T-shirt).

- Use Internet trend sites. With the widespread use of the Internet, lots of information about fashion is available to everyone. As with the fashion services like Promostyl, there are also a number of Internet professional services that one can subscribe to, but these are expensive and might not be worth the cost for most small store owners. Many other sites, however, will give you free but limited information. Trial and error is the best method for finding the ones that pertain to you, but as with all Internet information, you must pick through those sites that are just opinion-based and find the ones that offer solid,

Figure 6.5 *Fashion magazines help readers keep up-to-date with styles and trends.*

educated information from reliable, professional sources. One of our favorites is www.style.com, but you can find plenty of them with search words like "fashion" or "fashion trends."

- Take note of the media: television, sports, movies, athletes, and music stars all have a profound influence on what we wear. We see them and we want to emulate their glamour. Even *People* magazine has a section every week with the latest celebrity fashion and makeup trends. Rita Nakouzi, trend forecaster for

Promostyl, says that a part of her job is to attend places where fashion can be seen on the street: hot new restaurants and bars, concerts, museum events, clubs (not a bad job, right?).

- Shop the competition. Your neighboring competitors can give you excellent information about what kinds of looks are being accepted by the consumers in your area and which ones fall flat. If you are unsure about how quickly your potential clientele are willing to accept new fashions and colors, watch the markdown racks in similar stores in your area. These stores do not necessarily have to carry the exact merchandise as you for you to learn something about your customer's needs; watching the acceptance of certain price points, colors, and forward styles in shoes can give the apparel store owner an understanding of how his or her customer wants to shop. Combine the knowledge acquired through research and observation with the information that you have listed about your own target customer, and you will have a better understanding of their level of acceptance of new fads and trends.
- Watch everybody. Fashion, if you love it, becomes second nature. Without realizing that they are doing it, "fashionistas" constantly observe fashion trends everywhere they go. One of the best sources for fashion trend information is "the street." Particularly in large fashion-oriented cities like New York City, Los Angeles, Montreal, London, and Paris, you can get some of the best trend and fad information just by watching what fashion trendsetters that you see on the street are wearing—and how they are putting it together. You can observe all of this on visits that you make to market or even when you are on vacation, but fashion innovators and fashion influentials (people who are leaders within their own social circles) also live, work, and go to school in towns around the country like Richmond, Virginia. Be active in your community; watch how the society women of your town dress and where they go. Attend many of the same social events that they attend, and you will get a better understanding of their apparel needs.
- Learn from your own customers. Once your store is open, the best

information that you can get about what is right for your store is to listen to your customers. If many of them are asking for products or sizes that you don't offer or are shunning certain brand names, colors, or types of styles, then you need to pay attention. No matter how much you love a particular item or group of items that you have purchased for your store, it won't matter at all unless your customer likes it also.

The Buying Trip

Planning is key to the success of any market or trade show trip that is intended to result in the purchase of merchandise for your store. Earlier we discussed some sources for the purchase of goods. There are many ways that a store owner can find products to sell, but some of the primary ones are as follows:

Figure 6.6 *The MAGIC trade show features fashion products from all over the world and helps owners find new merchandise to sell.*

- Visit trade shows such as the MAGIC show (abbreviation for Men's Apparel Guild in California; the short name has stuck, but the meaning is no longer relevant since the show is held in Las Vegas, covers fashion for both men and women, and includes suppliers and buyers from all over the country and the world) in either February or August (Figure 6.6).
- Take market trips to areas that are key to your product lines. New York City ***Market weeks*** are excellent occasions to see the new lines that brands and wholesalers are offering for a particular season (Figure 6.7). Designated market weeks vary for different product lines: sportswear markets have set weeks every year, but they are different from the market weeks for lingerie, accessories, shoes, and other categories. You may choose to visit a different city

Figure 6.7 *A visit to New York City, especially during market weeks, is an excellent way to learn about new lines.*

to make your purchases, and there are trade shows and planned market events in many places around the country and around the world. *Women's Wear Daily* runs a special section (usually in December) with a calendar of dates and locations of market weeks, trade shows, and apparel fairs around the country.

- Once you have established a relationship with a branded supplier, you may be able to buy merchandise right in your store if that vendor has sales representatives that regularly visit local stores.
- Some wholesalers offer product through Internet Web sites, but you must first establish contact with these sources and identify yourself as a retailer. You would probably not buy from the same sites that retail customers use to order product.
- You may want to devote a portion of your store and a section of your purchasing budget for product from local crafters, artisans, or designers (Figure 6.8). Sometimes you can even work out a situation where the artisan will house product in your store on consignment with an arrangement that you only pay him or her when the consumer makes a purchase. If your store is large enough, you might even consider leasing out a section for compatible product with a deal that allows you to collect a portion of the proceeds from any sale of goods but absolves you from the risk of using your funds to purchase the inventory.

With the availability of ways to purchase merchandise right in your store or town, is it necessary to spend the time and money to travel to a trade show or market? We believe that the answer is yes. You cannot possibly see all the available offerings by just staying in your store or even in town. Traveling to a trade show like MAGIC or to New York City for a market week will allow you to see many vendors to compare lines of merchandise, to visit fashion-trendy stores in locations like SoHo to look for and observe fashion trends, and to review the lines of new sources. There is also no substitute for personal meetings and interactions with vendors in order to establish strong working relationships that bring some of those perks listed above later in this step.

Figure 6.8 *Products from local crafters, artisans, or designers can add a unique twist to a store.*

Plan before You Go

If you intend to visit a location away from your store to buy inventory, you need to make certain that you have completed an assortment plan such as the one in Table 6.1. Do your fashion trend research ahead of time and prepare a booklet with names and even pictures of merchandise that you feel is right for your store. If you are going to a trade show or a market week, decide well ahead of the trip which key vendors you wish to see, and make appointments with them. Particularly during market weeks, the vendors with the hottest products and name brands are typically busy, and as a new retailer, you won't be able to just show up and hope to see the line.

As discussed in Step 5 in the preparation of the Fixed Expenses List, travel is expensive and the store owner/buyer will want to evaluate each trip for its potential benefits in relationship to costs. You have already done an exercise for the preparation of your Fixed Expenses List in which you estimated the costs of your buying trips along with any entertainment expenses. It is important to plan early for these trips in order to get the lowest costs on air fare, hotel rooms, and so on. Don't compromise safety and convenience by just booking the lowest-priced hotel, particularly in a city like New York, without doing some research on the neighborhood and its proximity to market. You do not, however, need to stay at the most expensive hotel in town or fly first-class.

A new store owner must always take identification on a first trip to a new supplier: printed checks or a tax identification number to establish credit. You might have to pay in advance or COD (cash on delivery) until you have established credit and a relationship with the vendor.

Get to Know Your Suppliers

A word about vendor relationships: It is advisable to select a few key vendors with whom you will want to focus your purchases over a number of seasons. In planning your purchases, you might even establish a portion of your budget for a category for an important vendor (Nike, for example, if you have an athletic shoe store). While you will want to continually look at new lines and brands as they gain importance on the fashion scene, build strong relationships with a few key vendors. You will never have the buying power of Wal-Mart; however, you can achieve preferential treatment from suppliers if you show them that you plan to do significant business with them over a period of time. These benefits include the following:

1. *Better payment terms.* Instead of prepaying or paying COD, you can establish payment terms that allow you to pay after the goods are shipped to you from the vendor warehouse. Most likely, you can establish terms with your key suppliers that will allow you about 20 to 30 days after goods have shipped to pay your bills. When we complete the cash flow statement, you will be able to see how beneficial this can be in helping you to meet monthly expenses. If the goods are shipped via a method that allows fast receipt of goods into your store and if the merchandise is a hot seller, you might even be able to sell most of it before you have to pay for it. This is every retailer's ideal.
2. *Advertising support.* Most vendors have budgeted funds to provide marketing and advertising support to their good customers. Don't be afraid to ask for money to support your efforts to promote the vendor's product. There will be rules about this, and the vendor will probably only at best split the cost with you, but do inquire.
3. *Return privileges.* It is a given that you will return damaged or late goods to the supplier, but your key suppliers might also allow you on occasion to return merchandise in good condition that your customers show little to no interest in purchasing. This is a perk that is usually only offered to large retailers like Federated or Target, but even a small store can sometimes get assistance with slow sellers—or at least help in transferring the goods to another store that is selling them. The same applies to markdown allowances. Large retailers demand them, but sometimes a small store can get some monetary assistance in moving particularly slow-selling merchandise.

4. *Faster delivery times.* It is a "rule" of buying and selling that good customers get preference on first deliveries.
5. *Notification of special buys.* The vendor typically offers these when he is cleaning out a group of inventory or cutting up excess piece goods into items for quick selling. These special buy groups can be very beneficial to the store owner as a way to build sales and margin dollars and still be able to sell product at a below typical retail price for a special store event.
6. *Personal selling.* Advice from vendors on what other stores are selling, which items are hot fashion trends, and early preference on wanted items can be very valuable to the entrepreneur. If the supplier has local sales representatives, you can probably set up regular times when one will visit you to show new products and to help with old merchandise.

At this point in the planning process, you need to make a list of key suppliers with whom you plan to make the bulk of your purchases. Each entry on the list should show vendor name, key contact person, headquarters or showroom address, telephone and e-mail information, and any other data that is key to the vendor information. Next to that vendor, list the key categories and retail price points that you will purchase from this vendor. This list will be included in your business plan right after your Opening Inventory Plan (Table 6.2). To see an example of the way this list might look, consult the sample business plan in the Appendix. There is no right or wrong number of entries for this list, but it should reflect only your key vendors, not companies that you will buy from infrequently.

One additional truth to remember about buying the right merchandise for your store: *Never sacrifice quality for price.* Regardless of the price point, all customers expect value for the money that they spend. Most consumers today are very savvy when it comes to being able to distinguish between the offerings of many different retail outlets. More and more customers are using the

Internet to compare prices and style, so it is hard to pass off poor-quality merchandise at any price. While quality and value are important for every retailer, it is especially important to the small retailer that customers are happy with their purchases, even after they have worn and used them for a while. A small store depends upon the same loyal customers to shop at the store on a regular basis, so you want to make sure that they have many incentives to come back again and again. As the opening quote to this paragraph states, it is not enough to offer fashion if you don't also offer good quality.

Managing Assortments

There are many ways that a smart manager can maximize profits in his or her store, but the main way is to control expenses. Granted, it requires juggling ability to be able to negotiate the fine line between expenses that are necessary to create the kind of store where your target customer will want to shop and those that are extraneous. It takes time, some trial and error, and constant monitoring to manage that goal.

Controlling Markups and Markdowns

We have talked about markup and how important both regular (initial) markup and also maintained markup (the final markup after all reductions are taken) are to the financial health of the business. You have established an average initial markup for each of your categories, but that does not mean that every item in those categories must reflect the same initial markup. One of the main ways that a merchant can control maintained markup is through the wise use of markdowns.

It has already been established that not everything can be sold at regular price and that ***markdowns*** are inevitable in any business. Managing markdowns can, however, make all the difference between a profitable and an unprofitable season.

There are two types of markdowns in any retail apparel operation:

1. *Promotional markdowns.* When any merchandise is marked lower than the initial selling price for a specified length of time, and then taken back to original retail when that event has passed. Examples might be a store anniversary sale, a holiday event like "Black Friday" (the day after Thanksgiving), or a special sale that has been organized by the local Merchant's Association.
2. *Permanent or red-line markdowns.* Used to clear away merchandise at the end of a selling season or merchandise within season that has shown itself to be undesirable or a slow seller. An example of this would be the markdowns that typically occur in January on heavy winter coats and flannel sleepwear.

Planning for promotional markdowns usually occurs early in the season and perhaps even while you are writing orders for new merchandise. Perhaps your supplier has agreed to help you to offset the cost of these temporary markdowns by selling you extra stock on strong-selling items at a reduced price for your event. While you can never guess exactly what these markdowns will cost you, you can estimate a pretty good number after you have had some experience with these events. Planning for these events is crucial to the profitability of the store; they help to draw business during slow times or introduce new customers to your store. However, too many of them at the wrong times erode margins unnecessarily. Staging promotional events is a key element of your marketing plan.

Permanent markdowns are quite another thing, and most retailers shudder at the thought of having to take them. They do serve a purpose, however, in that they allow the owner to clean out unwanted goods in order to free up money to purchase newer, fresher, more fashion-right product. There are several important considerations to understand about taking permanent markdowns:

- All merchandise that is sold above the cost price still adds margin dollars to your treasury, even if they are not the margin dollars that you planned. Taking these markdowns in a well-planned and timely fashion will maximize the margin dollars that can be realized.
- There is an old adage in the industry: Your first markdown is the best and cheapest. Don't hold prices past the point when you know that they need adjustment. The longer you wait, the lower the price that you will have to take to move the inventory. Customers typically vote early on whether or not merchandise is desirable. Don't use well-worn excuses like "the weather isn't right" or "they will buy it next month" to allow you to hold undesirable goods. As soon as you realize that an item or group is not selling, take your first price reduction.
- Make an action plan for your permanent markdowns. By observing what seems to work for other similar stores, decide how soon after receipt of goods that you plan to take the first markdown and what percent off you will take, do the same for the second and third markdowns, and investigate options for clearing away the last pieces of your style or group. There are several options for this, for example, a sidewalk sale in a slow-selling period like the week of July 4 is one of them.
- Don't hold seasonal merchandise to the end of the selling season if it proves to be a slow-seller. Take action within the season to minimize margin erosion. "Remember, sometimes fast quarters are better than slow dollars" (Arrington, *Planning Inventory the Right Way*).
- Don't get caught up in the sale game that has hurt many department stores. Use your off-price events and permanent markdowns wisely, and don't allow the customer to think of your store as an off-price merchant. Avoiding gimmicky sales and excessive markdowns will help you to establish credibility with your customers as a quality, fashionable merchant (Dion and Topping, p. 54).
- Markdowns are a key element of the six-month buying plan process. Factoring

them into this planning tool will help the buyer/owner to see the effects on turnover, increases in open-to-buy, and positive cashflow. As well, he or she can see the negative effect on total sales, gross margin, and net profit.

Simulation

1. Complete the Opening Inventory Plan (Table 6.2), which is also on the CD-ROM
2. Prepare justifications for this form.
3. Make a list of Key Vendors with whom you will do business. Be prepared to place that list in your business plan next to the Opening Inventory Plan.
4. Complete Part V Product and Service of your business plan. In your plan you should include the three items listed above in this simulation. You should also include a discussion of the following:
    - Your planned assortment, key vendors, retail price points
    - Why your customers will come to you to buy your product (niche)
    - What trend research will make your store successful, and how you will keep your assortment current and up-to-date
    - How and where you will purchase inventory (goods) for the store.
5. Make notes of your merchandising plans for your store:
    - Promotional events
    - Possible fixture needs
    - Other methods for displaying your merchandise
6. You will need all of this information when you prepare later sections of the plan. You do *not* need to include this with Part V of your final business plan.

STEP 7
Marketing Plan

"We are limited not by our abilities, but by our vision."

—Unknown

PROJECT OBJECTIVES

This step in the simulation will guide you through:

- Evaluating the competition and conducting a market feasibility study
- Finding your place in the local market
- Anticipating customer habits/preferences and establishing a fine-tuned target customer base
- Clarifying your place in the product life cycle
- Formulating a customer service plan that works for your store
- Establishing retail prices
- Planning price breaks: promotions, clearance, customer incentives
- Branding your business
- Creating an effective promotional and advertising plan
- Analyzing cost/value relationships of advertising choices

This step in the preparation of a business plan requires the entrepreneur/student to formulate a marketing plan for the new store. Aside from financial planning, this is probably the most important part of the plan, as it addresses the steps that the owner will take in order to garner recognition for his new store and to draw the target customer to the store. This usually does not happen automatically. Certainly, a good location will aid in attracting customers, and many of them may stop by the store as they are shopping the mall or neighborhood in which the store is located. It is not enough, however, to depend on curious shoppers who just happen upon the store. The goal of the marketing campaign for a new store needs to be, first, to let as many target customers know at least once that a new store is opening and what the store will offer and, second, to build a base of loyal customers who will provide the main support for the business. Most small stores depend largely upon loyal, repeat customers to sustain them. Yes, they continually seek and hope for new and even temporary customers (tourists, for example), but there must be a solid base of frequent, repeat customers in order for the business to be strong.

Marketing Plan

The marketing plan is typically divided into four parts. In the simulation instructions at the end of this step, you will be told to develop a sound, thorough marketing plan. This plan should be made up of the elements in the following sections.

Target Customer

In the simulation instructions at the end of Step 1, you were assigned to write this first section of the marketing plan: to address the demographics, psychographics, and other pertinent information about your target customers, and to address the potential for your product categories in your particular local market. This step required you to consult many different sources of data, such as the Census Bureau statistics and other sources of local demographic data.

You should have discussed such factors about your customer as age ranges, income level, marital status, number of children, educational level, and type of job. This data is not just to make interesting reading; it is important for the store owner to understand the purchasing ability of the target customer, as well as what demographic factors will lead him or her to buy certain products or price points over others. The entrepreneur needs to understand this data before opening a store, as it would not be advisable, for example, to place a store selling luxury products in a blue-collar town or area where customers have limited disposable income and needs for such product.

You were also directed to research psychographic information about the target customer, such as lifestyle, preferences, values, hobbies, and any other relevant information that would help you to better understand what factors motivate them to purchase apparel. If you discovered, for example, that the majority of the potential customers for your store were also deeply religious, you would not want to open a store selling seductive apparel and you would need to consider this fact when planning advertising messages, in-store visuals, and even store events (such as perhaps no alcoholic beverages at the grand opening party).

Included in this section of the plan should also be a statement about the growth potential for your product lines and store. Are you dealing with merchandise that represents fads, trends, or classics? If your products are likely to become unwanted or go out of style in a few years, how will you keep your merchandise assortment updated? As your target customer base ages, will there be new consumers behind them to turn into potential customers? Will your current customers still want your products as they age?

You now have the opportunity to review that report and add to it as you finalize it for the Marketing Plan section of your business plan. Now that you have completed simulation assignments for other parts of the plan, you may have a different perspective on your customer and want to make some changes to the target customer profile.

As for format, you may elect to present this material as a chart, as a table, as a graphic, or just as a listing of data. You do not, however, want to just insert copied data from a source such as the Census Bureau Web site and expect the reader to interpret it. Instead, interpret the data for your audience (financial backers, potential clients, and others), using the data from your research as source and reference. Always keep in mind that you want everything in your business plan to be concise, appropriate, and reader-friendly. Don't bore the audience with rambling prose, and don't ask them to put information together and interpret it themselves. Get to the point as quickly as possible and back up your proposals with research and pertinent information.

Competition

This portion of the marketing plan should address the impact of other sources where your target customer can and might purchase similar products. It would be ideal to think that all new stores open and operate with completely different merchandise than anything being offered in the surrounding area, and thus do not impact the businesses of any other stores around them. In almost all cases, however, this is far from reality. Most of the time, a new store will depend at least in part on "stealing" retail sales from some competitors. There is, after all, a limit to the disposable income of your target customer and your mission is to make sure that your store is the first place she thinks of when she wants to purchase your lines of product.

It is crucial to both long-term and short-term success that the store owner correctly assesses the strengths and weaknesses of key competitors and stays informed over time on this subject. It would be very naive of any potential owner to state that she has no competition, or in the analysis of competitors to state that everything they do is wrong, while she will have all the right answers and all the right moves. Obviously, any competitor who has been in business more than a couple of years is doing something right, or that business would not have been able to survive in today's tough marketplace. A realistic analysis of key competitors would include the following:

- Location of competitor with proximity to the proposed new store
- Evaluation of their target customer base versus that planned for the new store.
- An assessment of their strengths and weaknesses. Compare those to the plans for the new store. Address the ways in which the new store will take advantage of any weaknesses in local competition.
- A recap of the ways in which competitors communicate to their customers: marketing message and methods, customer service, operational policies (return privileges, merchandise holds, methods of payment, availability of alterations, and others). Included with this should be an assessment of the new store's plans.
- Comparisons of how the product selection and product categories will differ in the new store with those of the key competitors; comparisons could be on styles, breadth of assortment, fashion timeliness of selection, sizes, or other pertinent factors.

This assessment should be made of any key competitors (also called ***direct competitors***) that the new owner feels are the major places where his target customer currently spends her apparel budget on similar product. In addition, the entrepreneur should also write a brief assessment of any ***indirect competitors*** who may impact the business. Typically, direct competitors are located in the same shopping area or at least the same town as the location for the new store; indirect competitors may also be in the same locale, but in today's widened shopping choices, they might even be on the Internet. If you plan to open a trendy athletic shoe store, for example, you may have a direct competitor in your town who sells similar product, you may have an indirect competitor like Foot Locker who does not sell the same product but can be a place where your target customer has previously been spending some of his athletic shoe budget, and you may also have a competitor whom you consider to be either direct or indirect who sells these kinds of shoes on a Web site or even through word of mouth. There is a fine line between direct and indirect. Only you as the potential owner can make the determination about how much of a threat another retail outlet is to your ability to generate sales revenue in your product categories.

Here's a major truth: Never underestimate the competition. Many a smart entrepreneur has let pride and ego keep him from a correct assessment of the threats from competition, both from those in business for a long time and from the new kids on the block. Keep your eyes open and your ear to the ground when it comes to the competition.

Pricing and Sales

Regardless of the strength of the product selection, no new store will succeed unless the retail prices of the items are competitive with similar product being offered in the marketplace. Although it is rare that an item or brand or product line appears on the fashion scene that is so compelling customers will pay any price to own it, we all have seen that happen. That is a great situation for a while until the fickle fashion customer moves on to the next exciting innovation, and then prices suddenly start to drop dramatically on the previous frontrunner. For a business venture in which the entrepreneur hopes to have long-term success, it is best to price product sensibly and realistically from the beginning to ensure that customers will be with the store for the long haul. We in no way mean to imply that the store owner should not expect to get as high a markup with resulting profit dollars as is feasible, but we encourage the entrepreneur to refrain from greed. Consumers will only be taken advantage of for so long before they rebel and move on.

How to know the right retail prices to attach to the product in question? The first answer to that is *research*. Your previous assessment of competition will give you the most important information about where retail prices should be. It is an old apparel industry adage that says that the buyer (owner) does not set the retail prices; the consumer does. It is important to know what savvy customers are willing to pay for their fashion merchandise, and retail prices on similar merchandise that is selling well in similar stores will be your best gauge for that. Your vendor partners will also be able to give you valuable insight into the right retail prices to put on your merchandise. While it is against fair trade laws for retail stores to act in ***collusion*** to set retail prices and for vendors to force retailers to offer product at a predetermined selling price, many brands will attempt to "strongly suggest" that you use a certain retail price on hot brands and products. You may find that some brands will not even sell product to your store if they have an exclusive relationship with another similar store in your town or area, or they may only sell certain items or lines to you that are not currently being offered by other retailers in your trading area.

Your retail prices will be impacted greatly by the markup goals that you have set for yourself. After completing all the financial documents for your store and realizing what kind of gross margin dollars and markup percents you will need to achieve in order to pay your fixed expenses and still deliver a profit, you may have to make some adjustments to your retail price points. Keeping in mind that industry adage about consumers setting the retail prices, you may have to take a multitiered approach to retail pricing and markups. The ability to do this is partly the result of good product research and not buying the first items or brands shown to you. As you build your assortment plan, you will begin to see the items that can deliver higher-margin dollars and those that are so restricted because of cost, brand pressure, or competitive store exposure that they must carry a lower margin. Previously, we talked about the fact that you should have, in each category, some items that you will carry at all times in your store that will make up the base dollars for their category, and that are generally sought by your target customer. These are the items that can potentially offer the highest gross margin percents. There are several reasons for this: You probably will not have to mark them down as often as your high-fashion items, you will sell them in larger quantities and may be able to negotiate a slightly lower price from your suppliers, and you can shop around and be very wise about where you initially purchase them, as they are likely to be items that are available from several different sources. Mega retailers like Wal-Mart, Target, JCPenney, Federated Stores, and the like have huge buying power and are able to negotiate lower prices on many of their day-in, day-out items and thus realize higher-margin dollars on these classics. Even a small store, however, has resources, and you should shop around, negotiate, and be wise about these margin builder items and classifications.

Set Retail Price Points

The new store owner must make a decision early in the planning process about the retail price points that she will offer. There are several ways to approach this. She can elect to use even retail prices only: $20, $50, $100. She can elect to use uneven retail prices only: $22, $52, $102. She can elect to use cent endings: $19.99, $49.99, $99.99. She can also choose to mix it up and use all of the above strategies, but that is a bad idea. Customers have a gut reaction to price points and often categorize a store by the way that they set their retail price points; .99 endings, for example, often denote "discount" or "bargain basement" stores. The decisions about pricing that the owner makes will largely depend upon the store image that she is trying to project. Make the image decisions early and let every other decision about the store revolve around and complement that.

Limit Your Retail Price Points

It is also advisable to keep your price points for each category within a certain range, have a limited number of price points that you use, and make sure that each category's prices complement those of merchandise in the other categories. In our simulated store, for example, we would not want to offer pants at price points everywhere from $14.99 to $400 because we would only confuse our customers, and we have found that the $14.99 shopper would be frightened off by the $400 pants, and the $400 shopper would think that our quality is inferior if she saw the $14.99 pants. We would also not choose to offer $400 pants and $9.99 tops. Again, customer confusion and anger would occur. You would not likely see most $400 pant customers wearing a $9.99 top as part of the outfit.

For stores that offer higher prices or even mid-priced merchandise with regular retails that end in zero, the owner will often elect to use another ending to denote sale prices. When the $400 pants are marked down at the end of the season (red-lined), the owner might choose to mark them down to 25 percent off, but use an ending like $299 or even a .99 ending when the price gets to the lowest markdown. Again, that is an individual store decision, but it is less confusing to the regular customers if the manager keeps the same pricing strategy consistent throughout the store and for each sale event. Again, each decision about store operations and store policies impacts the overall store image.

Set Pricing Guidelines

It is wise for the new store owner to set down some guidelines before the store opens concerning retail pricing strategies. This policy should be part of your marketing plan and should be addressed in your business plan. The strategy should cover the following points:

1. Initial retail price points: What will be the policy for assigning price points on new merchandise? What endings will they carry? What is the range of pricing within a particular category? Here you should refer back to your opening assortment plan (Step 6) to see how your average price points are planned per category; you have already made some important decisions about that question.
2. What markups will you seek per category? To answer this question, go back to your Sales and Gross Margin Exercise (Step 5) to see what you planned to achieve both in regular-priced product (initial markup) and on-sale markups (combination of promotional and permanent markdowns). Make some statements for the marketing plan that deal with the necessity for these markup goals and how you plan to achieve them.

3. Retail price point strategy for on-sale goods. When you run promotional events, how often and when will they occur? Do you plan to put all categories of merchandise on sale? Only selected categories? Only certain merchandise within a category? Will this differ from promotional event to promotional event? You will be asked to address these questions in more detail in the simulation instructions for the marketing calendar that you will prepare to be included with your marketing plan.

You should similarly address permanent markdowns. When will you take these? At season end? When an item is determined to be a "dog"? As with your other pricing strategies, decide at what point you will take markdown action on a slow seller. For example, in our store we have decided to give a new item or group about four to six weeks of regular-priced selling before we take our first markdown, which will in most cases be 20 percent off. For any pieces that do not sell at that first markdown within two weeks, we will further reduce the price to 35 percent off original retail. After two more weeks of sales activities, we will attempt to sell the remaining units at 50 percent off. If any stock is left after that, we will either offer it at our sidewalk sale or donate it to charity. We covered all of this in Step 5 as you were creating assumptions to calculate your sales and gross margin chart, so you have probably made these decisions. You now need to clarify them in the Pricing section of the marketing plan.

What other creative approaches will the store use in setting price points? Will there be multiple pricing such as one might find on the panty table at a store like Victoria's Secret, where a single pair of panties is something like $6, but one can buy five pairs for $25?

Consider Customer Service Strategies

The final area of importance for the Pricing section of the plan is a discussion of the customer service strategies that the store will offer, along with an enumeration of the costs associated with each service. These services may fall under categories such as free alterations, or even alterations that the customer pays for. The exact policy for such services should be detailed, along with the methods for achieving the desired level of service. For alterations, a store owner may choose to have an in-house permanent seamstress to handle these, or he may choose to contract with a firm or individual to perform these services at an outside location. In either case, the cost of the service must be stated. Almost all stores today must offer several forms of payment: Will the new store accept cash, checks, charge cards, debit cards? Which ones will the store accept and what will be the attendant charge for that service? Stores must pay the MasterCard Company or bank for the privilege of using their products. In this manner, all customer service practices must be detailed and costed.

Advertising and Promotion

In this final section of the marketing plan, the student/entrepreneur should address in detail his plans for promoting and advertising his store and product line. By this point, having identified the store niche and formulated product categories, price points, and customer service plans, you should be prepared to articulate in a concise manner exactly what you want to tell the public about your business. You should then describe how you will get this message across to your target audience, with a listing of advertising and promotional vehicles that you intend to use and why. It is important to note frequency of use for each method and any cost associated with each vehicle that you intend to use. These costs should then be tabulated into a budget for pre-opening marketing and advertising and one for annual first-year expenses. Pre-opening or start-up expenses are those that you intend to incur before the store is open, so preopening advertising and marketing costs should be those associated with announcing the opening of your store and getting your message to your target audience at least one time.

Note: The following segment represents marketing advice from Michael A. Sisti of Sisti and Others, Inc. Michael is a branding, marketing, and advertising consultant with 50 years' experience in the marketing communications field. During that career, he has provided services for some of the largest corporations in the world, as well as many sole practitioner start-ups.

Branding

One of the most misunderstood yet critical aspects of building a successful business is branding. It is every bit as important for a small business as it is for Ralph Lauren or Neiman Marcus.

By definition, ***branding*** is the philosophical and emotional attributes of a company, its products, and its services. It is essentially all the intangible aspects of the company, the perception that becomes the company's image.

Branding starts with selecting a name for the business. Considerations for the selection process should be a name that is memorable and relates to the product, service, or key benefit of the company. Company names that encompass the name of the owner or partners are extremely popular and show a commitment by the founder. Among the types of names that are not recommended are those made up of initials. These are usually difficult to remember, adding to the marketing costs of maintaining top-of-mind awareness. In selecting a name, review the names used by your competitors. This will prevent you from choosing a name that is the same or similar to one of them. Hair salons often make this mistake. For some unknown reason, they nearly always create a name that is a play on words involving hair. And this makes none of them stand out and be remembered.

Building a successful brand requires the development of a foundation document (see Figure 7.1). This consists of six elements: mission statement, vision statement, company values, brand position, brand personality, and brand promise. An introspective examination of purpose for starting the company, your vision for the company, the type of customer you want to attract, and your style of doing business will provide clues for these elements. Once this document is created, every action your company takes, regardless of how small or insignificant, should be measured against this brand foundation. If an activity doesn't support it, you should change that activity to fit the brand. All employee behavior and every customer touchpoint, including answering the phone and greeting visitors, all become part of the branding process. Problem solving and customer service are particularly important in building a positive brand experience.

In addition to the behavioral aspects of supporting the brand image, it is important that all marketing communications, signage, in-store design, merchandising, and so on reflect the brand. These elements strongly impact the brand per-

Rose Knows Clothes
Brand Foundation

Mission
We are in business to sell fashionable clothing and accessories that are not found in the mainstream shops and department stores.

Vision
Our success will rely on our ability to identify and represent small and upcoming manufacturers who produce lines of clothing, shoes, and accessories with distinctive designs.

Values
We value our ability to recognize new fashion trends, upcoming designers, and manufacturers with a unique flair. We value our customers who are savvy shoppers, willing to pay a reasonable but not outlandish premium for unique, quality merchandise. We value the extraordinary experience we provide every customer that visits our store.

Brand Personality
The personality of our brand matches the excitement and exuberance a person experiences when they find something really special. We are fun, adventurous, and sophisticated.

Brand Position
We position ourselves as an upscale boutique, delivering unique fashion merchandise, coupled with the highest levels of customer service.

Brand Promise
We promise a satisfying shopping experience, where our customers will always find something unusual and out of the ordinary.

Figure 7.1 *All branding starts with a brand foundation. The elements of the foundation describe the vision of the founders and what the brand aspires to be.*

ception. The ***logo***, stationery, signage, and other elements that constitute the brand identity should always be displayed in a consistent fashion (see Figure 7.2). To accomplish this, develop a set of simple brand standards. These rules will guide the design and structure of the brand identity. The standards will also determine the usage of the logo and company colors, where appropriate in all communications efforts, such as advertising, collateral material, Web site, newsletters, and so on. And most importantly, it will guide the look, feel, tone, style, and personality projected by these communications. (See the Rose Knows Clothes simulated store folder on the CD-ROM for branding and advertising examples.)

Figure 7.2 *Stationery and business cards should support elements of the brand foundation. It is important to choose fonts and illustrations that convey the store's image.*

Successful branding, even for the smallest company, will convert shoppers into customers and make them loyal advocates of your business. It will cause them to propagate your marketing message and expand your brand's influence. It is the essence of word-of-mouth advertising. Psychologically, people like to boast about making a particular purchase or discovering a new resource. By convincing a friend or relative to make the same purchase or use the same outlet, their purchase decision and their shopping know-how has been reinforced.

Advertising/Promoting the Business

"Build a better mousetrap and people will beat a path to your door." While this may in fact be true, if those people don't know you have a mousetrap that is better, or where they can buy one, the path to your door will go unused. Communicating this information is necessary to sell your products. And this marketing communications effort carries a substantial cost. You should therefore budget for it, and it must produce results that justify the cost. This can happen only if the process is managed properly.

With advertising, you can create messages using graphics, dramatic sounds, and compelling words that communicate your message, create awareness, and build demand for your products and services (see Figure 7.3). You also want to select the appropriate media that complements your message. For example, *Popular Mechanics* would not be the best publication in which to advertise seductive lingerie. A better choice would be an upscale lifestyle magazine or the fashion section of a newspaper.

Advertising offers many venues for delivering your message. Traditional media include newspaper, magazine, radio, television, yellow page directories, and out-of-home advertising (see Figure 7.4). With the explosive growth of the Web, Internet advertising has quickly become a major component of most companies' advertising mix. Web site banners, pop-up messages, and broadcast

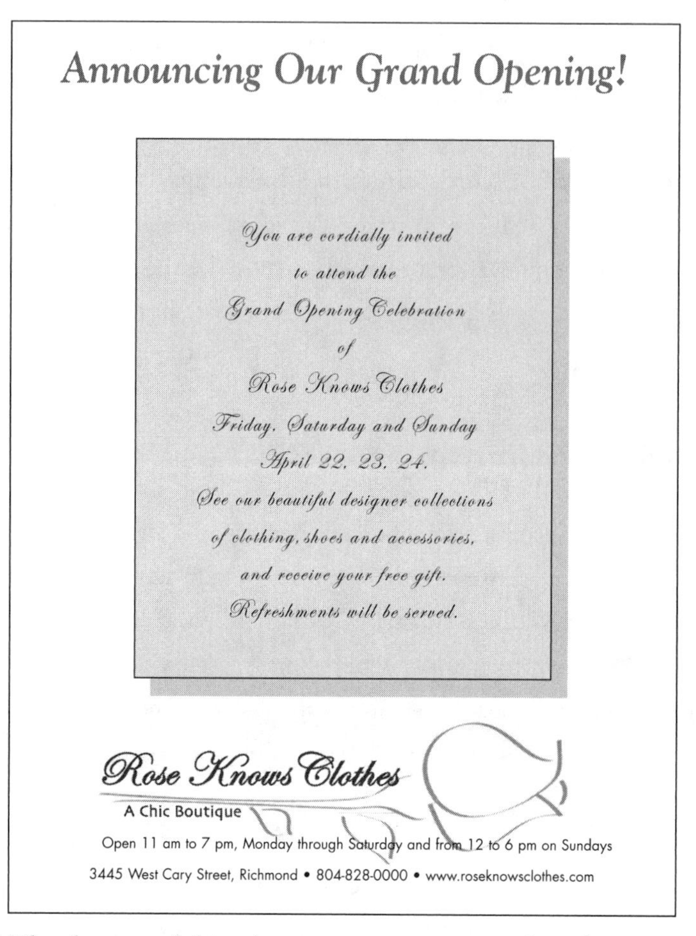

Figure 7.3 *The elegance of this ad creates an expectation of quality. It gets the attention of the audience, while projecting and enhancing the brand.*

e-mails have become more sophisticated and more effective, offering a cost-efficient alternative or addition to a communications program.

Advertising vs. Publicity

The most expensive communications discipline is advertising, but it is also likely to generate the best results. With advertising, you pay the media companies for

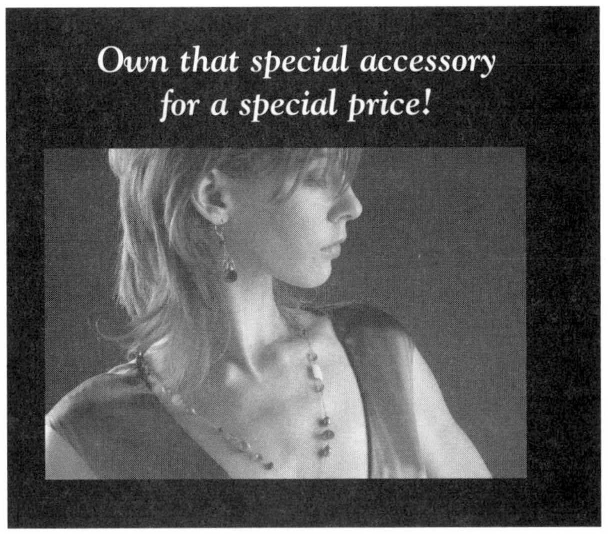

Figure 7.4 *This ad looks nothing like a typical sales ad, yet it gets the attention of the audience with a dramatic photo. It delivers a strong branding message and advertises the merchandise.*

the space (print) or the time (radio, television) and you control the message. You decide the style, tone, wording, and so on of the message, as well as how frequently your message will be delivered.

Following advertising, publicity is the next most active method of delivering marketing communications. Unlike advertising, there is no cost for a message to appear in a publication or on air. However, there is also little or no control over the message. Preparing a "release" or providing the media with a story or material for publication requires careful planning and well-crafted written material (see Figure 7.5). It also entails telephone or e-mail follow-up to promote the story, answer questions, or clear up any ambiguities by the reporter or editor. Once in the hands of the press, however, the material may be interpreted, edited, and sometimes distorted into something different than you intended. But when

Rose Knows Clothes
A Chic Boutique
3445 West Cary Street, Richmond, VA 23220 • 804-828-0000 • WWW.ROSEKNOWSCLOTHES.COM

For Release on:
April 8

Contact:
Rose Regni

From Monday, April 12 through Saturday April 24, Rose Knows Clothes will hold a special sale on all accessories, shoes and assorted merchandise. All items will be reduced from 40% to 70%.

Featured in the sale will be the exclusive jewelry and accessory lines from Italian designer, Elisa Orsini. This will be a one-time opportunity to get selected items from this coveted collection at 70% off.

To help us celebrate this event, Orsini's US representative, Anna Giovanielli will be visiting Richmond and will be on hand Friday, the 23rd from 1 PM to 6 PM and on Saturday the 24th, from 10 AM to 1 PM. Ms. Giovanielli will preview the new Fall lines and she will be offering some sample merchandise, not regularly sold through retail.

This is an extraordinary event for Richmond and a wonderful opportunity for our customers to purchase exclusive and unique merchandise at low prices.

For more information visit our website at www.roseknowsclothes.com, or call us at 804-828-0000.

Note: Color and black and white photography of select products enclosed.

Figure 7.5 *A press release delivers news immediately, including the release date, contact person, and details of the news. It is succinct and may provide photography.*

a positive story or report is published, there is an implied third-party endorsement from the publication or station, which adds credibility to your story.

The cost of publicity is the time it takes to put together the story or release, plus any photography or other documentation that accompanies the package. If a public relations consultant is utilized to create the material and follow up with the media, those costs will become part of the cost of the published story. Compared to advertising, publicity carries a fraction of the cost and can produce very beneficial results.

Other Media Considerations

Whether for advertising or publicity, choose your media carefully, in order to save time and money. All media outlets provide information on their costs and their audiences. These ***media kits*** offer information for you to make an informed decision when considering which publications to use for your marketing. Compare costs against audience size to partially determine value, but also look at the demographic information provided. A publication that has a low cost per thousand readers is a bargain for your needs only if its audience fits your target market in terms of gender, income, education, geography, and so on. Radio stations' audiences vary not only by the interest in the station's offering (type of music, talk radio, all news, all sports, etc.) but also by the time of day. Morning and evening "drive time" will skew an audience to a higher percentage of businesspeople, while "day parts" will often include more homemakers. With television, various time slots will also attract different audiences, but so will the different shows. A sporting event and a soap opera on the same station will draw completely different audiences. Weighing all these factors contribute to your selection of appropriate media. Your goal is to choose the right mix of media that reaches your largest target audience for the lowest cost. This applies to both advertising and publicity. There is no point in pursuing a story in a publication whose audience has nothing in common with your products and services.

Beyond traditional media outlets for advertising and publicity, there are many other choices for delivering marketing messages. Out-of-home advertising, which comprises roadside billboards, transportation terminal and shelter posters, movie theater trailers, and so on, can be very effective, but these are often expensive forms of advertising compared to traditional advertising. Direct mail is another communications vehicle that should be considered part of a marketing program. Using carefully selected mail lists, a mailing can be targeted to your ideal prospect. There are also low-budget mailings that group many advertisers into one mailing. These come in the form of coupon mailings or packages containing flyers from several companies. While these mailers are attractively priced, results vary due to the dilution of your message among the many other advertisers.

"Yellow page" or telephone directory is another advertising venue that often deserves consideration within a communications program. For certain business categories, particularly emergency services, it is indispensable. However, it can be expensive, particularly if it doesn't generate phone calls, and it requires a commitment for a year, without cancellation. It is best to research the effectiveness of directory advertising in your specific business and carefully weigh the cost against the anticipated results.

The Internet has become an indispensable tool for marketing communications. Nearly everyone has become comfortable, if not expert at using its power to research products, locate resources, and make purchases. There are several ways to effectively use the Internet. First, you should have a Web site, whether for informational purposes or to actually sell merchandise. If you opt to sell product over the Internet, your site will have to be more complex, with ordering and shopping features, plus payment software. In addition, the site will require constant maintenance to upgrade and change product offerings, pricing, and so on. A site that sells products must also be heavily promoted itself, in order to drive the necessary traffic to it. This requires optimization for search engines and the use of banners on other sites to develop click-through traffic. In essence, it becomes a business unto itself.

If you decide not to sell merchandise on the Internet, a Web site is still an important tool. Use it to display your products, promote your benefits, and project your brand. Provide contact information plus directions and maps to your location. Also, frequently update the site with announcements about sales, new merchandise arrivals, and information about trends and developments in your field. By constantly adding or changing information, visitors will return to the site on a regular basis (see Figure 7.6).

The other powerful Internet tool is e-mail. To be effective here, you must quickly develop e-mail lists to maintain a communication link with your

Figure 7.6 *A Web site is an important tool for promoting a business. It should be inviting, provide information, and encourage the visitor to want to learn more.*

customers and prospects. Even before you open your business, start collecting e-mails of friends and relatives. Through them, collect addresses of their friends and family and build as large a list as you can. Use that list to announce your grand opening, and invite everyone on your list to a grand-opening event. If that's not practical, invite them to visit your place of business and get a free gift for stopping by. Once you are open for business, start collecting e-mail addresses by offering to provide advance notice of sales or promotions. Do this with both your in-store and online visitors. Carefully crafted e-mail messages and flyers sent to a growing list of e-mail recipients can provide great results without any expense. One word of caution: Do not oversolicit your list. If the frequency gets annoying, you will lose your audience.

Creating the Right Plan

Analyzing the above information is one part of creating an effective media plan. The other factor is budget. The challenge is determining the minimum amount of money required to deliver the results that will meet or exceed the stated goal. This is a very difficult task, as so many factors affect the result. The number of competitors and the effectiveness of their marketing efforts, for instance, will impact the success of your plan.

Once the budget is established, the process of selecting the frequency and mix of media takes place. Depending on the budget allocation, it is always best to select at least three different types of media to distribute the message. Frequency will also be affected by the amount of budget available. Effective marketing communication is a building process that requires a combination of repetition and message delivery from different sources to penetrate the clutter of advertising your prospects receive on a daily basis. Therefore, if you run an ad just once, you should not expect much of a response. Increasing frequency will improve your response rate geometrically.

A key part of your communications plan is a calendar. Once you've decided what media to use and what frequency makes sense, develop a calendar of when the communications will run. If your business is seasonal, that will have a major impact as to when you increase or decrease your frequency. A typical calendar will show the insertions in monthly publications, daily or weekly newspapers, radio and TV schedules, planned press release and feature story dates, mailing drops, and so on. Also include your e-mail broadcasts in the calendar, so you will have a complete picture of your marketing effort at a glance.

Because of the complexity of creating a plan, it is advised to seek the help of a competent advertising agency. If you choose this direction, consider agencies that handle clients with similar budgets to yours and have experience in your specific field. An advertising agency can also help you create effective advertising and improve your chances for success. With the high stakes of advertising, it is best to use the expertise of professionals in this endeavor. That being said, small retailers and other commercial businesses, particularly start-ups, may not be able to afford the services of an advertising or public relations agency. Or their budgets may be too small for an agency to consider. As an alternative, there are consultants and freelancers who can be engaged on a project basis to assist in the development of a program. Most media companies, newspapers, magazines, radio and TV stations, and so on will offer creative services to small businesses for a nominal fee. This can be very cost-effective, but the resulting advertising will probably lack a creative edge. For assistance with developing a media plan, a knowledgeable media representative from a newspaper or local radio station often will be willing to help a small business like yours with the selection of media outlets and frequency, even those that he or she does not represent.

The other key factor to consider when writing a plan is that it must be dynamic. The market is constantly changing, affecting your business in many ways. If your sales exceed goal, you may want to increase your budget as a way to reinvest the additional profit. Or you may encounter unexpected competition, or a smaller market from which to draw customers than you originally projected. All of these factors and more will force you to be continuously reviewing and adjusting your budget and your plan.

Developing the Marketing Communications Message

The key axiom in crafting and delivering a marketing message is that every communiqué must be single-focused. This is true whether the message is delivered through a letter, advertisement, publicity release, or in any other form. Whatever discipline is used to deliver a message, remember to create it within the confines of the brand standards. For this reason, most companies develop a format for their advertising and printed material, using the same colors, type fonts, and basic layout throughout. This makes the communications easily recognizable and helps build awareness. But for some unexplained reason, when sale ads or flyers are created, these standards are often ignored and a communication is produced that is cluttered, contains multiple messages, and has no relation to any previous material distributed by the company. And for this reason, communications of this type underdeliver in their response rates. See Figure 7.7 for an example of an ineffective sales ad. There is no reason that a sale ad can't be produced using the brand standards and still garner outstanding results.

In creating messages, it is important to reflect on the audience's perspective. What do they want to hear? What will motivate them to take action? In too many instances, business owners are talking to themselves in their advertising: *I have wonderful merchandise. I am a savvy merchandise buyer.* Instead, the message should offer a benefit to the target: *You would look wonderful in our new designer collection. We've gone to great lengths to make this merchandise affordable for you.*

Figure 7.7 *This ad has no flow, emphasizes too many points, and is an example of not supporting the brand foundation.*

Creativity can have a dramatic effect on the ability of your message to break through the clutter and resonate with your prospect. Delivering your communication in an ingenious way allows your message to stick with far less repetition than an ordinary message presented in a dull way. And always leave a little something to the imagination. Don't give out all the details in any given communication. Raise questions. When people "connect the dots," they connect with the message. Consumers appreciate creativity, humor, beauty, and art, so consider utilizing these elements when developing advertising messages. After all, most advertising appears in entertainment venues, where people expect to be entertained. They are offended when you intrude with messages that are dull or unpleasant or insult their intelligence.

Just about any marketing communication will generate results, unless it is so bad that it has a negative effect. Successful marketing occurs when you achieve maximum results from minimal expenditures. To make that happen, follow the steps outlined above. Simply deliver a creatively composed, single message that is customer-focused and is wrapped within your brand standards.

Analyzing Cost/Value Relationships of Communications Programs

The cost of marketing is an investment in the company in the same way as the cost of inventory, equipment, fixtures, and so on. And like any investment, you should be looking for a return on that investment. Once you have established a budget and begun your communications program, carefully track the sales that come directly and indirectly from that effort. Comparing the investment in marketing against the profit on sales will provide you with the actual return on your investment. At the beginning of the campaign, your costs will exceed your profit. This will occur until the marketing message begins to resonate with your audience and they respond by making purchases.

An important component of implementing a communications program is testing its effectiveness. The myriad of variables that impact a campaign make it difficult to accurately project results. In the very beginning of any new initiative, it is important to test your message, as well as the media that has been selected. If the budget allows, focus groups and informal surveys can assist in determining the best message to deliver. As an alternative, you can try different messages in different media. Of course, it will also be necessary to compare the results of different media outlets, factoring in their costs. Here again, once you have produced a successful campaign, utilizing an effective message and a good media mix, use this benchmark to test future campaigns. Constant tweaking of the message and the media may enable a continuously improving return on investment.

It is also important to measure the effect your communications program has on the brand perception. This can be determined by surveys and inquiries of customers. Do a simple survey using a postcard or a pop-up survey on your Web site when you first launch your campaign. The results of this initial poll will establish a benchmark to measure against future surveys. In this way you will see the results change, hopefully with a better brand image over time. This goodwill and positive brand perception that emerges from your marketing communications have a value that cannot be measured in sales or profits. It does, however, contribute to the ***net worth*** of your company and becomes a key factor in valuing or selling your company. The other benefit of this branding is how it contributes to future sales through lower marketing expense.

Selecting a Name for Your Store

Keeping in mind the advice given above by Mr. Sisti, you now need to select a name for your store. The selection of a name is very personal and something that you will have to live with for quite some time. Therefore, you want a name that will stand the test of time. You will spend a lot of time and money on branding the store name and image, so be certain that you (and any partners) are comfortable with the name long before the sign goes up over the door.

For our simulated store, we have chosen a name that uses the first name of one of the owners, combined that with a rhyming, kitschy slogan that we hope that our target customer will be able to easily remember and one that denotes our primary business. The name is *Rose Knows Clothes*. The name lends itself easily to the creation of a popular logo in the world of fashion: the name super-imposed across a long-stemmed pink rose—all on a pink (feminine) back-ground. You may want a more sophisticated or serious name. That is entirely your choice, but it is important at this point to decide upon a name, as you will need it to create your logo and marketing campaign.

To illustrate one way to use the brand image and consistent message throughout, we have created a logo for our simulated store, *Rose Knows Clothes*. It is a rosebud with the store named printed across the stem. Earlier in this step, we showed you a business card and letterhead, which we will use for customer communication; both use our logo and color scheme. (See color images on the CD-ROM in the Rose Knows Clothes simulated store folder.) We want to train the customer to recognize our colors and logo whenever we use them in advertising. We must now make certain that our image, service level, and products continue to offer consistency and live up to customer expectations.

Note: In an Internet-driven market, it is important to select a store name that is easy to search for on the Web. Try to avoid overused names or names that contain unusual punctuation.

Creating a Marketing Calendar

In the interest of making your business plan and Marketing/Advertising section easy to read and understand, you should create a calendar showing when and how you intend to market your business and what that effort will cost. In Step 5, you established a marketing and advertising first-year annual budget for your store when you prepared your list of fixed expenses. You must now show how you will use the bulk of that budget and also a specific list of expenses that you intend to incur for your pre-opening marketing campaign. Your calendar can be created in any program you wish, such as Microsoft Word, Excel, or even a design program. Others like to write their plans on an actual yearly calendar, and still others prefer to just use a simple list to show promotional and marketing events and costs.

For our simulated store, we allotted a budget of $13,200 of first-year fixed expenses for marketing and advertising (budget represents 4 percent of annual planned sales revenue of $330,000). Our marketing calendar might look something like the one shown in Table 7.1. (See a blank version of the marketing calendar on the CD-ROM.) We have not, at this point, accounted for every dollar that we intend to spend during the first year, as we want to leave some dollars open for opportunities that arise during the year and we realize that we can make changes to our plan after the store is open. We have, however, shown how we intend to spend the bulk of our budget.

TABLE 7.1 MARKETING CALENDAR

PROMOTIONAL EVENT	PREOPENING	MONTH(S) PLANNED	TOTAL COSTS
Creation of Web site	$400	March	$400
Press kit	$150	March	$150
2,500 direct-mail postcards with postage costs	$900	March	$900
1 week radio ad on local station	$400	April	$400
1/2-page ad in *Style Weekly*	$300	April	$400
Grand-opening party, food and music	$800	April	$800
Internet broadcast e-mail	0	March	0
TOTAL OPENING COSTS			**$3,050**
Style Weekly 1/4-page ads		May, September, November, December, February, April	$4,000
RVA Magazine 1/2-page ad		April, August, November, February	$800
Press kits		August, November, April	$150
Direct-mail postcards		July, September, November, January, April	$4,500
Richmond Times newspaper ads, 1/4 page		May, September, December, March	$500
Internet broadcast e-mail		May, September, December	0
Update of Web site		July, November	$200
TOTAL ANNUAL EXPENSES			**$10,150**

Marketing Ideas for Small Stores

As stated earlier in this step, the entrepreneur does not have to spend a lot of money to market the store; she just needs to be creative. Table 7.2 outlines some clever ideas that small stores have used to entice customers to buy.

TABLE 7.2 MARKETING IDEAS FOR SMALL STORES

EVENT	ORIGINATOR	METHOD
Step Sale	Beth Ferris, owner, The Complete Horse, Richmond, Virginia	Uses for in-store sales. Sets up three areas (tables or racks) for sale merchandise: the first table the customer sees has a sign shaped like a shoe print and is marked as Step 1 where merchandise is marked 10% below original retail, the second rack is marked Step 2 and is 20% off, and the third table is Step 3 and is marked at 30% off original retail price. It is eye-catching and clever.
30% off on 30 items for 30 days	Beth Ferris, owner, The Complete Horse, Richmond, Virginia	Beth picks one slow sales month each year (often June or July is a slow month for retailers of apparel) and plans for it by creating a calendar that she puts in the bags of any customers who shop in her store for the month or two before the sale. On the calendar, she indicates a different item in her store for each day that she is open in a month. Each item is on sale for that one day only for 30 percent off original retail price.
The old shoe promotion	Client of Sharon Brooks & Assoc. ad agency	Used old shoes strategically around town with messages attached to them about a special event. Advantage: eye-catching and different. Can be any item that is different and of low value and that draws attention.
Multi-store fashion show	Heidi Story fashions, Que Bella Shoes, and other stores in Carytown	Work with other noncompeting stores in your shopping area to put on a fashion show in a local restaurant. Find a restaurant with a separate space and host the show during lunchtime to attract women to show new season's merchandise. Cooperative advertising for event.
Secret Sale/ Flashbulb Sale	Andrea Walters, VCU student project idea	Four times a year, advertise a random "secret sale," without divulging the time during the month when the discount will be given. Instead, program an alarm or flashbulb of some kind to go off at a certain time during the store hours on one day of the month. When the buzzer sounds, all customers who are in the store will get 50% off anything that they purchase for a period of 15 minutes.
Host a party after hours	Henry Shoes, Turnstyle Clothing, and other stores on Broad St., Richmond, Virginia	During art show on Broad Street (once a month in summer on Friday nights), stores along the walk stay open late, offer light refreshments, and some have music or even DJs in store. If your town doesn't have these events, create one with other neighboring stores.

EVENT	ORIGINATOR	METHOD
Sponsor or create a contest or local event	This has been used by many small stores in many different product areas.	Find out where your customers like to go and what their interests are and then sponsor or help to sponsor a charity event, sporting event, or other venue where your target customers will likely be.
T-shirts and shopping bag giveaways	Many stores have done this as a marketing method.	Especially good for new stores who want to get their name out in the public eye. Design or purchase a desirable giveaway item for the first 100 customers who come into the store. A T-shirt or nice canvas tote bag (which are desirable enough that the customers will want to use them) is a way to get free publicity for your store.
Offer seminar on subject of interest to your customer	This has been used successfully by several different kinds of stores, kitchen and home goods stores especially.	Host an event and invite customers to a seminar to teach them about your products (gift and home goods stores), have models in the store wearing new fashions (clothing stores), bring in a bra-fitting specialist (lingerie stores).
Cooperative discounts with noncompeting stores	This works for stores that draw the same customer but sell different products.	In-store advertising: display advertising brochures for another business such as nail salon, spa, shoe store, or restaurant. Brochure or card could offer discount on products from the cooperative business. That business could also post a brochure or card for your business that also offers discounts to customers who bring in the card or brochure when they shop.
Web site offer	Advice from Sharon Brooks of SB&A advertising agency, Richmond	Every page of your Website should have a way to capture names and information about potential customers. Offer a benefit (discount, free something) in exchange for names and addresses.
Trade-offs	Advice from Sharon Brooks	Make a trade with local venue for marketing: offer merchandise or prizes to local radio or cable station in exchange for mention of your store.
Best customer specials	Advice from Sharon Brooks	Incentives for best customers: invitations to private events held after hours; best customers offered early look at new products; discounts or other incentives or gifts given.

Simulation

1. Write your complete marketing and advertising plan, which will be Part VII as outlined in the Introduction. Include the four parts as described at the beginning of this step.
2. Include budgets for both pre-opening costs and for annual first-year expenses.
3. Create a calendar that shows when these marketing and promotional events will take place.

STEP 8
Personnel and Sales Management Plan

"It's not the employer who pays the wages. Employers only handle the money. It's the customer who pays the wages."

—Henry Ford

PROJECT OBJECTIVES

This step in the simulation will guide you through:

- Planning personnel needs
- Finding the right people
- Determining salaries and benefits
- Managing people effectively

Personnel Needs

Most businesspeople that we have met seem to think that they are indispensable and that their businesses would totally fall apart if they were not around. That is true to a certain extent because every organization needs a leader to guide it and give it direction. However, every organization also needs followers who will take the guidance and direction and put it into action. Both types are necessary, and one will not succeed without the other.

Smart business entrepreneurs will realize early on that they cannot be all things to all people and trying to do everything will usually result in doing nothing. Many try and most fail because they end up wearing themselves out and wasting the capital they have expended. Knowing your limitations and deciding what you can do and what you cannot do will pay off in the long run.

How do you go about evaluating your personnel needs? Start by listing the things that the owner does not do well, duties that he or she does not have the time to do, and/or jobs that need to be done simultaneously. The list will form the basis for determining personnel needs (Figure 8.1). Remember, you cannot do it all by yourself. Trust us, you can't. The sooner you realize this, the better off you will be.

The business plan needs to address how many people the business will require to meet its projections. That number will obviously vary depending on the size of the business, but even if it is only two people—the owner and one

Figure 8.1 *Create a list of strengths and weaknesses to evaluate your personnel needs.*

employee—they must be accounted for in the business plan. In addition, all of the expenses related to personnel must be included in the business plan.

Owners must also assess the many different types of employees available to them. Will a part-time after-hours student meet your needs, or do you need a full-time eight-hour-a-day person who is multitalented and has a Ph.D.? The types of employees you can hire vary greatly, and choosing the right type is as important as how many you need—maybe even more so.

In an employers guide written for the Portland Community College Office of Students with Disabilities (http://spot.pcc.edu/~rjacobs/career/types_of_employees.htm), Robin Jacobs defines the following six types of employees that a business needs to consider:

- Full-time employees
- Part-time employees
- Temporary employees
- Leased employees
- Job share employees
- Employees with co-employers

Full-Time Employees

Workers considered full-time employees usually work 40 hours per week. Full-time employees most often receive employee benefits that may include vacation time, retirement fund contributions, health benefits, sick leave, and other benefits provided by the employer.

Part-Time Employees

Employees who work considerably less than 40 hours per week are considered part-time workers. While some employee benefits may be provided, benefits are usually quite limited or reduced in proportion to the amount of time worked.

Temporary Employees

Temporary workers, often called "temp" employees, are workers employed by a temporary service business. Employees usually work for a short period of time at different companies to which they are assigned.

The workers are employees of the temporary business, not the companies where the work is performed. The temporary service pays workers' wages and withholds taxes, Social Security, unemployment insurance, and workers' compensation from paychecks like other employers do. Some temporary businesses, but very few, offer benefit programs—health insurance, retirement plans, paid vacation, and sick leave.

The types of workers most commonly hired by temporary services include office and clerical support staff, technical workers, and professionals—doctors, lawyers, and corporate executives. A small retail store can benefit from this kind of employee because it does not need full-time bookkeepers, custodial staff, or accountants. Students who like to work a few hours after school or on weekends also would fit into this category.

Temporary work arrangements often attract workers who desire work schedule flexibility, an opportunity to check out potential employers, and a means of acquiring work experience and contacts for getting a foot in the door at a desired company.

Leased Employees

Leased employees are employed by service firms that supply workers to client companies on a temporary basis. Leased employees may be assigned to one job for perhaps a year or longer. Leased workers are hired and paid by the lease service firm, not the client companies where the work is performed.

Job Share Employees

In a job share arrangement, two or more employees share one full-time job. For example, two employees might agree to work 20 hours each per week. Benefits

provided by the employer are prorated by share. Many employers are now offering job share options to help retain employees and increase worker satisfaction. This option is good for small stores and would fit the lifestyle of current stay-at-home moms/dads who also want some income but cannot work 40-hour weeks.

Employees with Co-Employers

This is a relatively new but growing trend among small businesses wishing to turn over human resource responsibilities to ***professional employer organizations (PEOs)*** to manage and handle. In this situation, employees work for the small business, but a separate company, the PEO, manages all the human resource (HR) responsibilities, such as:

- Paychecks
- Benefits
- The handling of personnel issues
- Tax withholdings
- Retirement programs
- Unemployment insurance
- Health insurance
- Workers' compensation

Organizational Structure and Organizational Chart

An organizational structure will help the owner define his or her relationship to the business and its employees. It will define who in the business has the authority and responsibility to accomplish the business's goals. It will also define who in the business has the authority to assign tasks to employees as well as direct them in how the tasks are to be accomplished.

The visual component of the organization's structure is the organizational chart. It readily shows who has what authority, who reports to whom, and

where business tasks are accomplished. It is a visual illustration of the organizational structure. Figure 8.2 is a simple organizational chart that is used by many small businesses.

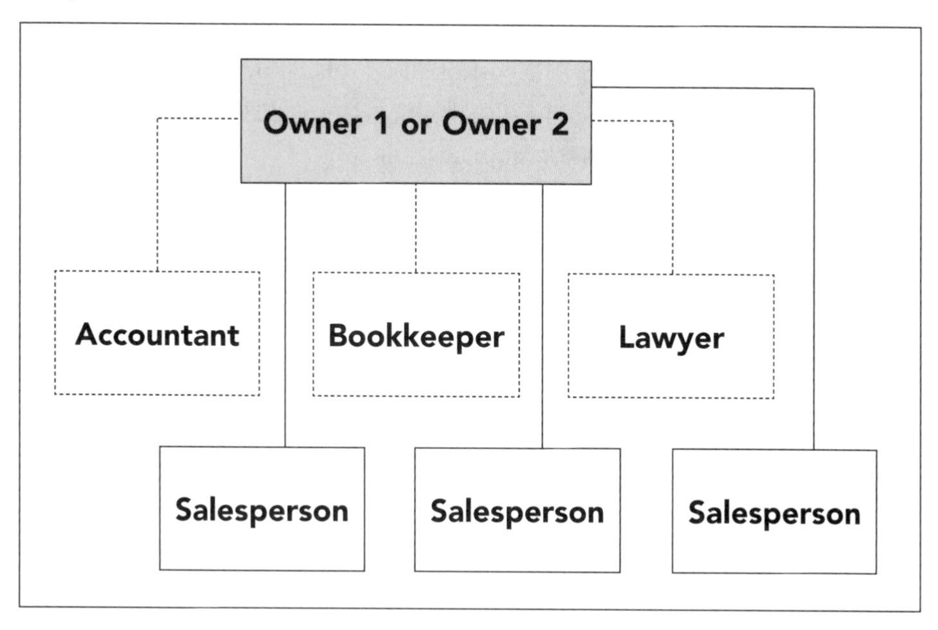

Figure 8.2 *Organizational chart for Rose Knows Clothes.*

A good organizational chart will be simple yet readily describe the power/reporting structure in the business. At a glance, the chart will define the size of the business and number of employees. It will also show the duties for which employees are responsible, to whom employees report, who the managers are, and who the owner/chief executive is.

Finding the Right Employees

Knowing how many employees a business needs and knowing what they will need is one thing. Finding the appropriate person to fill those needs is a totally different thing. Having the most skilled person in the world to sell your product means nothing if you and that person do not get along or if his or her personality clashes with your other employees.

In another sense, having a nice, well-mannered, customer-oriented person means nothing if that person cannot add 1 + 1 or does not know how to report to work on time. Naturally, in the best of all worlds, your employees will not only be compatible with you and each other but they will also be conscientious, intelligent, loyal, knowledgeable, and work for free. (Actually, we are kidding about the free part.)

Finding the right employee begins with knowing what you want him or her to do upon being hired. You will need to write job descriptions for each type of job you want filled. These descriptions are very important because once you hire someone, the terms under which that person is hired are legally binding. Of course, most employers cover themselves by including in job descriptions the dreaded (by most employees) phrase "Other duties as assigned."

Employers can find lots of help in writing job descriptions. Trade associations, unions, educational institutions, business development organizations, and employment search firms all have publications that can be beneficial in preparing a job description. The job description needs to detail all of the duties expected of the employee as well as the skills the potential employee will need to bring to the job.

An excellent resource online for developing job descriptions is BLR's Web site at www.hr.blr.com. The site has thousands of job descriptions available for review and use by business owners. It permits those seeking employees to use already-developed descriptions or modify them to the particular use desired. The site also provides information on all aspects of human resource management and can be very helpful to the novice employer. A sample sales associate position job description can be found in the sample business plan in the Appendix.

However, knowing that you need employees, how many, and what you want them to do once they are hired is only half the battle. You need also, of course, to find people who are actually interested in applying for the jobs you need to fill. Most of you reading this text have probably never been on the hiring side of the employment equation. In fact, most of you have only been on the job-seeking side.

Figure 8.3 *Assess the communication tools you will need to have a successful operation.*

The communication explosion of today is no doubt the friend of both the employer and the job seeker because, in the past, the only way to let the public know you were hiring was an ad in the local newspaper. The same holds true for those seeking employment. Today we have the Internet, instant messaging, e-mail, cell phones, thousand-channel cable networks, satellite networks, iPhones, iPods, BlackBerries, and on and on (Figure 8.3). A person who wants to avoid constant communication would have to live on Mars and even then would have to put up with the occasional rover rolling by, sending data back to Earth.

All of these avenues of communication can be overwhelming, and it behooves the small business owner to carefully target the area from which job applicants are drawn. If you are only a two-person operation, your other employee is likely going to be the person with whom you went into business or someone else that you know. As the business expands, you can take advantage of all of the resources mentioned above. If you are only seeking part-time employees, your likely sources of employees are going to be students, retirees, stay-at-home moms/dads, friends, and relatives.

Interviewing and Hiring Employees

We would be less than candid if we did not tell you up front that interviewing and hiring employees is ultimately a crapshoot. You can do all the due diligence available, but in the long run, until the person actually performs on the job, the employer has no guarantee of how good the employee will be. A person who has the very best qualifications on paper and conducts him- or herself well during an interview may be a total bust. The reverse is also true. Someone you may have doubts about could turn into the very best person for the job. Unfortunately, employees do not come with written guarantees. The hiring official can only make sure that he or she learns as much as possible about the potential employee before a job is offered.

If an employer is lucky, he or she will have multiple applicants from which to choose. Most hiring officials will use the written job application to narrow down the field, and a final candidate will be chosen by interviewing the top applicants. Obviously, it is best if the interviews can be done in person.

In recent years, interviewing job applicants has become a minefield given our litigious society. When interviewing a potential employee, the interviewer has to be very careful about the kinds of questions that are asked. In addition, all applicants must be treated equally, and no preferential treatment is to be given to one candidate over another.

While very small businesses (those under 50 employees) usually do not have to worry about many of the very specific government regulations regarding hiring and employees, all employers must heed the rules regarding discrimination as to age, sex, and race. The Equal Employment Opportunity Commission is the federal agency that governs hiring practices. Employers can find all of the rules regarding hiring on the EEOC's Web site at www.eeoc.gov.

It is common knowledge that the person being interviewed for a job must prepare him- or herself in advance. The same advice applies to the job interviewer as well. In order for the interview process to be as successful as possible, the interviewer must do his or her homework and prepare just as diligently as the person being interviewed. It is insulting to the job applicant to come into a job interview and find that the interviewer is ill prepared. Not only will the applicant be disappointed but also the likelihood of a bad hire will be increased.

In the book *Managing for Dummies*, the authors provide excellent guidance on the preparation for the job interview from the interviewer side. Their very clear and concise help can be found at their Web site at http://www.dummies.com/WileyCDA/DummiesArticle/id-933,subcat-BUSINESS.html?print=true. The article on job interviewing covers such topics as preparing for the interview, asking the right questions, interviewing dos and don'ts, and evaluation of candidates.

If you have used the Internet to help you come up with interview questions, then the applicant has probably already accessed the same Web sites. So try to come up with some questions that the applicant cannot anticipate. Remember, you know a lot more about your business and the kind of employee you are looking for than the person being interviewed. Be sure to make the interview process a good one for both you and the applicant. Even if the applicant does not get the job, it is important that he or she leaves the interview and the application process with a positive view of you and your company.

Employee Benefits and Compensation

People—at least all the ones we know—work for one reason and one reason only: to sustain themselves either spiritually or physically. The only ones who can work to sustain their spirit are usually mooching off someone else or are independently wealthy. Everybody else works so that they can make a living in order to pay for a roof over their head and put food on the table. Those who can fulfill themselves both spiritually and physically in their jobs are indeed lucky.

As an employer it is in your best interest to have happy employees. Happy employees are productive employees, and productive employees lead to profitable businesses. Profitable businesses lead to happy employers. Happy employers make for a happy workplace, and a happy workplace usually leads to happy employees. It is a circle of business life.

In their book *Raving Fans*, Ken Blanchard and Sheldon Bowles offer numerous examples of how ecstatic employees take customer service to stratospheric levels. In addition to being great ambassadors for your company, happy employees:

- Are more likely to stay with the company
- Are absent fewer days than unhappy employees
- File fewer grievances
- Complete their work more quickly
- Produce higher-quality work
- Find ways to improve their effectiveness
- Share their enthusiasm with colleagues

So what makes a happy employee? According to Todd Raphael in an article published in *Workforce* in 2002:

Employees should have some degree of satisfaction with their jobs. Equally important, they should be energized, motivated, and eager to try something new. This is especially true for salespeople, upper management, creative employees, and individuals in other positions that demand a lot of energy. You might not want your accountant so eager for change that she decides to create a new kind of balance sheet. Your balance sheet is probably fine, especially if your employees are relatively content, but agitated enough to keep up with today's pace and need for change.

As you might expect, having a successful business is like walking a tightrope. You need to balance making sure your employees are happy along with what is the best for the business. You need to give them the tools and benefits they need to be successful at their work, but you also need to keep them on their toes so they do not become complacent.

The following commentary on employee benefits was gleaned from SBA publication PM-3, *Managing Employee Benefits*:

Why offer employees benefits? Here are some reasons:

- To attract and hold capable people
- To keep up with competition
- To foster good morale
- To keep employment channels open by providing opportunities for advancement and promotion

A combination of benefits programs is the most effective and efficient means of meeting economic security needs. For many employers, a benefit plan is an integral part of total compensation, because employers either pay the entire cost of a benefit plan or have employees contribute a portion of premium costs for their coverage.

Employers must provide certain legally mandated benefits and insurance coverage for their employees:

- Social Security
- Unemployment insurance
- Workers' compensation

The following is stated in the Social Security Administration's *Social Security Handbook*:

> The Social Security Act and related laws established a number of programs which have the basic objectives of providing for the material needs of individuals and families, protecting aged and disabled persons against the expense of illness that would otherwise exhaust their savings, keeping families together, and giving children the opportunity to grow up in health and security.

Funding for the Social Security program comes from payments by employers, employees, and self-employed persons into an insurance fund that provides income during retirement years. Full retirement benefits normally become available at age 65. Other aspects of Social Security deal with survivor, dependent, and disability benefits; Medicare; Supplemental Security Income; and Medicaid.

Unemployment insurance benefits are payable under the laws of individual states from the Federal-State Unemployment Compensation Program. Employer payments, based on total payroll, contribute to the program. Workers' compensation provides benefits to workers disabled by occupational illness or injury. Each state mandates coverage and provides benefits. In most states, private insurance or an employer self-insurance arrangement provides the coverage.

There are also optional benefits that an employer might provide, including:

- Health insurance
- Disability insurance
- Life insurance
- A retirement plan
- Leave/vacation

Broadly defined, a benefit plan can include other items such as bonuses, service awards, reimbursement of employee educational expenses, and perquisites appropriate to employee responsibility.

Before an employer can implement any benefit plan, he or she needs to ask him- or herself the following questions:

- What can the company afford?
- How much is the company willing to pay?
- Do you want employee input?
- What should the benefit plan accomplish?
- Is a good health insurance plan more important than a retirement plan?
- Do you want the company to administer the benefit plan, or do you want it done by an outside source?

Once these questions are answered, the employer can then go benefit shopping for his/her employees.

While all employees are usually eligible for benefits such as health and other insurance, retirement plans, and leave, key employees have come to expect certain other additional benefits related to their increased levels of responsibility. Following are some of the perquisites employers may want to consider for top performers and key employees:

- Extra vacation
- Spouse travel on business

- Professional memberships
- Professional publications
- Parking
- Tuition programs
- Dependent day care
- End-of-year bonus
- Merchandise discounts
- Holiday gifts
- Health club memberships

Like basic benefits, perquisites help attract and keep good employees. You can balance the far higher cost of providing some perquisites with expectations of increased production from the employees who benefit.

What should you pay your employees? Here again the Internet is your friend. There are countless Web sites that offer help in determining what your employees should be paid. Business.com alone lists 37 sites where you can look up salary information. An article titled "Guide to Setting Employee Pay Levels and Salaries" by John Henshell at the same Web site provides great guidance on determining employee salaries. Mr. Henshell's basic advice is to:

- Establish a pay structure.
- Review salary surveys.
- Make adjustments for your market and industry if necessary.
- Make adjustments based on the competitiveness of your benefits package.

Other items we would add to his list include the following:

- Research what your competitors are paying.
- Make sure you comply with any minimum wage laws in your state or city.

- Review classified ads in local newspapers and magazines.
- Know what you can afford to pay.
- Leave room in your salary structure for raises and bonuses.
- Make sure you comply with IRS and state government laws requiring tax withholding.

Effective People Management

Without a doubt, the number of books and articles on managing people effectively must number in the thousands, if not in the millions. The art of managing people has been studied, and studied, and studied, yet it still causes a great deal of consternation on both the part of managers and those who are managed. No one has yet to come up with the perfect way to deal with employees.

As director of an office containing 35 employees, I (the author) was constantly astonished by the many challenges presented to me by such a small group. The office had a budget of several million dollars and the responsibility of delivering all of a federal agency's programs to an entire state. However, the personnel aspects of the job took the most time.

Most of the employees were professional people, but I still spent a lot of time on personnel issues. I wrongly assumed that if employees were highly educated and well paid, they would be easy to supervise. It was a valuable lesson for me to learn to never underestimate the difficulties and complexities of personnel management.

An owner of a small business cannot be expected to spend inordinate amounts of time on personnel management. He or she is most often worrying about too many other things. Even so, it behooves the employer to establish some basic rules regarding the management of his or her employees and to write them down in an employee manual.

The employee manual does not need to be a long document, but by having

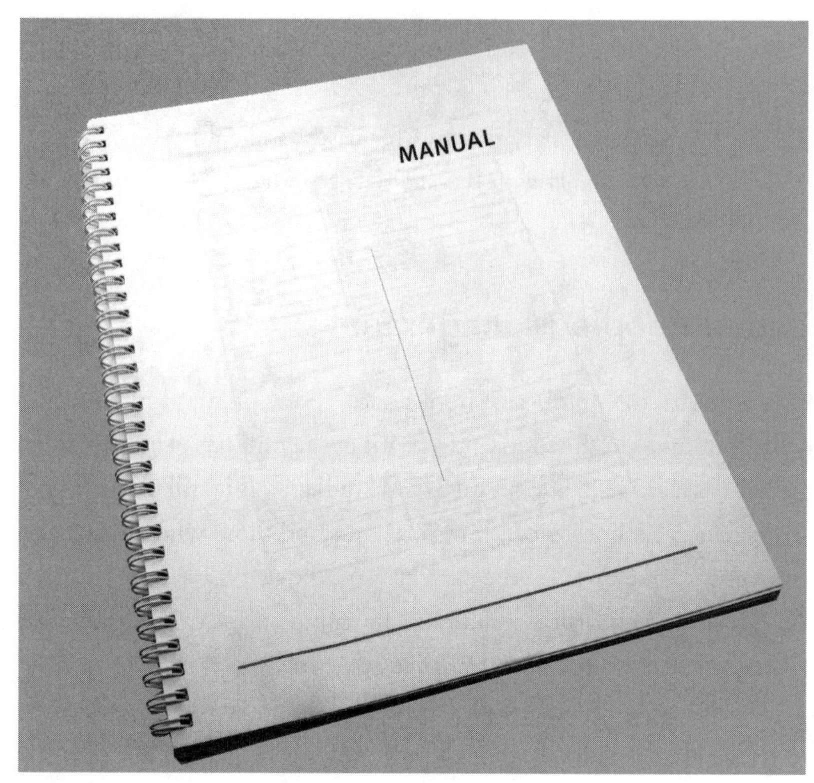

Figure 8.4 *It is important to have an employee manual to define basic procedures and rules.*

some simple rules in writing that all employees must follow, everyone will benefit (Figure 8.4). Not only does such a manual clearly define the rules of employment, it also does the following:

- Defines expectations from the employee
- Provides for consistency of application of the rules regarding employment.
- Reduces the danger of miscommunication
- Provides a positive way of maintaining management discretion while firmly communicating mutual expectations

- Defines how the employee's performance will be monitored
- Defines incentives and rewards the employee can expect
- Defines the firm's vacation and leave policy
- Describes any training the employee must undergo

When an employee is hired, the employer should meet with him or her and present the employee with a copy of the employee manual and explain its contents. At the end of the meeting, the employee should sign a document stating that the manual has been received and that the contents are understood.

Monitoring Employee Performance

There are four widely used monitors for the evaluation of employee performance. These principles are explained by Jim Dion and Ted Topping in *Start and Run a Retail Business* in their chapter on sales management. These principles can be very helpful to the new entrepreneur who is looking for ways to evaluate employees. Remember, even if you only have a few employees, you need to rate and evaluate them. These four principles are outlined in Table 8.1.

You might want to have a formal evaluation for long-standing employees once a year, at which time you will probably give them raises if performance warrants. You will certainly use that time to let them know how they are doing, and to give them suggestions for improvement, or for continued good performance. It is also a good time to talk to them about your plans for the business and where they might fit into those plans. It is wise to avoid overpromising, as that can sometimes cause legal problems, but you should encourage employees if there is reason for encouragement. If you plan to expand the business, for example, you might tell a highly valued employee that he or she is in consideration for advancement. Many studies on employee satisfaction have shown that a well-deserved pat on the back is just as important to a good employee as a raise.

TABLE 8.1 RATING EMPLOYEE PERFORMANCE			
RATIO	FORMULA	EXPLANATION	GOOD PERFORMANCE
1. Selling cost	Wages and benefits / Net sales	Employee annual salary and benefits is divided by the sales that they produce annually to show a percent performance ratio.	10% to 14% for a store whose focus is personalized service.
2. Conversion Rate	No. of transactions / No. of customers	Shows how good the sales associates are in converting shoppers in your store into purchasers.	Without being too "pushy," good sales associates should be able to convert about 40% to 50% of people who enter a store into buyers.
3. Average dollar value of transactions	$ Sales per week / No. of transactions per week	Demonstrates sales associate's ability to sell multiple items to same customer—very important for the success of small stores.	Owners must set expectations that relate to store price points and type of product.
4. Items per transaction	No. of items per week / No. of transactions per week	Similar to row 3, but shows number of items that sales associates are able to sell to each customer on average.	Same guidelines as in row 3.

We hope that you will not wait, however, for the once-a-year formal review to talk with and work with your employees to help them grow and develop and bring in more business for the store. Employees want your input, if it is handled in a positive manner, and they usually welcome your demonstrations of what works best with consumers. Have an on-the-job training program and keep them informed of new merchandise, upcoming promotions, new vendors or brands, and any changes in store policy that impact them or the customers that they serve. Give them the tools to do a good job and the odds are favorable that they will do their best. Some managers hold regular meetings to update and encourage employees. These can be done before the store opens, after the store closes, or at a separate store meeting (perhaps combined with a meal or small party). Of course, there are many ways to encourage and reward employees without spending a lot of money. Again, as we stated earlier, there are a myriad of books on this subject. We have outlined some of the more common ways in Table 8.2, but be creative. Come up with some new ideas that you know will be meaningful to your employees.

The benefits of valued employees cannot be overemphasized. Your employees are the most important assets that your business possesses. You must treat them with respect and dignity, and you must compensate them fairly. Your business cannot prosper without them.

Simulation

1. Prepare the Personnel section of your business plan. Refer to the outline in the Introduction, Part VIII. This section will address the following topics:
    a. The background and relevant experience of the owner(s). Include discussion of strengths and weaknesses; note how you will compensate for weaknesses (e.g., hire someone to do that task, or assign that task to the partner). If the business does not show enough profit in the first year to pay a living salary to the owner(s), address the issue of how you will support yourself.
    b. Details regarding employees—how many, who they will be, salaries, vacations and other policies, and benefits.
    c. Salary/benefit costs in relationship to sales revenue plan and fixed costs.
    d. Other employee perks and rewards.
2. Prepare an organizational chart for your business. Include the chart with your Personnel section. Include the support team of people who will work with you and advise you on key issues. Examples would be accountant, lawyer, financial partners, and perhaps others.
3. Write a sample job description for each different type of position in your business. Include these in your Personnel Plan section.
4. Prepare a sample weekly sales coverage chart showing how you will plan for

TABLE 8.2 EMPLOYEE REWARDS		
TYPE	**EXPLANATION**	**COST**
Smile and congratulations	Every employee wants this reward. And will usually be motivated to do more.	Time and effort.
Outside training	Good employees want to be able to perform their jobs well; keeping them up-to-date with skills will enhance their abilities.	Depends upon type of training; probably not as much as you think. Check with local colleges, training centers, your suppliers, and vendors.
Trip	Taking a valued employee with you when you buy products or go to markets is both training and a reward.	Airfare, meals, hotel room—depends upon location of market.
Discounts on store products	Three-fold benefit: Pleases employee, advertises product, educates employee in product attributes to pass on to customers.	Some loss of margin, but increase in product turnover. Set limits on the percent of the discount and number of products.
Free store merchandise	Same as above.	Higher cost as cost and profit on items are lost. Seek vendor help, limit occurrences.
Food	Employees like this reward, as it saves them time and money. Use it as part of store meeting/reward at holidays.	Can be relatively inexpensive. Pizza for lunch or bagels for breakfast usually work fine.
Gift certificate	Can be for a spa, restaurant, entertainment event, etc. Makes employees feel special. Can serve as contest prize for such things as top sales for month or year.	Can be expensive, but costs can be controlled depending upon chosen venue. Use these rewards sparingly and only for top employees. Can substitute for temporary delay of raise, smaller-than-expected raise, or delay in start of benefits.
Party	Good incentive to increase cooperation among employees.	You can control costs by choosing an inexpensive location (in the store or your home). Can limit to holidays or special events.

the right number of employees to be on the sales floor during the times that the store will be open. Include information about when the owner(s) will be in the store or when a manager will be present. If the owner or manager is expected to cover sales for a particular time period, show that on the chart. Use the chart to show salary costs for a "typical" week in your store, and extrapolate that to show annual salary expenses and percent of that cost to total planned sales. The business plan in the appendix shows an example of how a chart may be prepared. Include this with your Personnel section.

5. Include the personal financial statements and resumes that you prepared at the end of Step 4. You may either include them at the end of the Personnel section or at the end of the plan.

STEP 9
Visual Merchandising and Store Layout

"Vision without action is a daydream; action without vision is a nightmare." —*Japanese proverb*

PROJECT OBJECTIVES

This step in the simulation will guide you through:

- Preplanning considerations
- Planning store layout: matching product to store size and configuration
- Attracting the target customer through careful planning of inside and outside of store
- Smart planning of fixture layout and merchandise location
- Using effective display and unique store attributes to sell merchandise
- Merchandising tricks and styles, including in-store and window display

The quote at the beginning of this step clearly spells out what can happen to new store owners who start throwing fixtures and merchandise into their new stores without smart, deliberate planning. For a small boutique particularly, but certainly for all stores, ***visual merchandising***, or the planning and execution of fixture layout and merchandise placement, is crucial. From the moment that a customer walks up to the front of the store to the moment he or she walks out the door with purchase in hand, the customer will be greatly affected by the visual merchandising of the store. Think about some of your own shopping experiences and what made them good or bad, memorable or forgettable. Would you walk into a store to buy apparel if the windows were empty, dirty, or dated? Would you buy apparel from a store with all the fixtures placed haphazardly and the merchandise thrown on those fixtures in a completely unorganized way, possibly with garments half hanging off the hangers? Probably not, even if you had heard that the store carried some item that you thought that you really wanted to purchase. Maybe you would buy, but only if the price on the merchandise is marked at half off retail. As a new store owner, you can't afford to offer those prices, so the visual merchandising of your store needs to match the quality, fashionability, and status of your merchandise. Remember also, all customers are attracted by beauty and order, no matter what price they pay for the merchandise inside; however, luxury shoppers expect even more.

Preplanning Considerations

Before the new owner begins to lay out fixtures or plan merchandise placement, some considerations demand attention. Any store planning needs to start with the customer in mind: What will she expect to see or want to find in your store? What store image will most entice him to buy? It is very important that the store design project the image that the owner wants the store to communicate to the public. So, before you start planning, make sure that you have thought carefully about your image and your ***mission statement*** (more on mission statements in Step 12).

At the same time that you are evaluating what will grab your target consumer and inspire him or her to buy, you must also do some research to understand what requirements the federal, state, and local municipality will impose upon your new store.

Government Regulations

The Americans with Disabilities Act (ADA) was written into law in July 1990 to protect disabled persons with special needs when they require access to public accommodations. This law covers many types of public facilities and businesses, including retail stores. It mandates that these facilities provide access for both customers and employees with disabilities. New stores and remodeled facilities must pay particular attention to this law and its provisions. Provisions include handicapped parking access; availability of ramps if necessary; location and size of entrances, emergency exits, fitting rooms, and bathrooms; aisle widths; counter heights; elevators; and access to drinking fountains and public telephones. Further information is available from www.usdoj.gov/crt/ada/adahom1.htm.

You should also check state and municipal codes, as they are sometimes more strict than even the federal government or may have different standards to consider. If there are major renovations being made to an existing store, often the store must be brought up to code standards (Lopez, pp. 9–14).

Considerations for Our Simulated Store

For our simulated store, most of the renovations are being done by the owners of the building. As renters, we are only responsible for the things that we add to the inside of the store. As shown in the renovated floor plans in Step 2, our store is only one floor at street level, so we do not have to be concerned about an elevator. In addition, the bathroom was already built, so the dressing rooms are our major concern and we have made sure that they are to code. We realize that we must pay attention to aisle widths.

Store Layout

One of the decisions that the entrepreneur must make about his new store is whether or not to employ help in planning his interior layout and fixture placement. If he so chooses, there are a number of places that he can go to get help. The variables are how much he can or is willing to spend and his own level of expertise in the area of store planning. As with other considerations about opening the store, it is a decision that must be made carefully. If he lacks expertise in this area but still decides to do it himself, he may end up with a visual merchandising plan that actually detracts from the appeal of his product lines. He may end up with wasted space, the wrong kinds of fixtures for his merchandise, or just an unappealing color plan or decor or maybe the wrong kind of lighting plan to show off his merchandise. On the other hand, he can hurt his chances of making a profit in the first couple of years if he spends too much on this part of the project.

Getting Help with Store Planning

Where can you as the store owner go to find help if you decide that you need it? Following are some ideas:

- Hire a store planner. A store planner is an "individual . . . with specialized talents in presenting merchandise so that it will be purchased by customers" (Lopez, p. 1). A professional store planner will likely be the choice that will cost the most but will also be the one with the most experience and credentials.
- Hire a fixture consultant. The fixture company that provides your fixturing needs may have an in-house consultant who will help the owner plan the best use of the fixtures in the store. This service may not cost any extra, but the owner is, of course, obligated to buy all fixtures from one source.
- Find a student or college professor in the field of interior design to help. We did this for our simulated store, and we found Jennifer Hamilton, recent

graduate school student and now a professor in the Department of Interior Design at Virginia Commonwealth University. For about $1,000, she designed the renovations for our store that the owner agreed to make, and then planned a very attractive and workable layout for our fixtures, built-in shelving, and checkout area.

- Take some courses and learn more about visual merchandising and store planning so that you can do a lot of the work and planning yourself.

Matching Product Needs to Store Size and Configuration

The most important step in matching product with store came when you selected your store. At that time, you should have known what your product categories were to be and how well they would fit into the store that you selected to rent. Hopefully, you didn't try to open a furniture store in an 800-square-foot space or a lingerie store in a 10,000-square-foot facility. Of course, it is not likely that your store space is ideal for your product category or that it has all the amenities that you desire—unless, of course, you had it built to your specifications. So, in most cases, you will take an existing store space and mold it as best you can to act as a showcase for your product. Large department stores will probably have a visual merchandiser on staff—perhaps even a whole team of visual merchandisers who receive instructions and new season setup plans from the visual team in the corporate office. You will not have that luxury, so even after the initial store planning and setup stage is complete, you will either need to acquire or hone your own visual merchandising skills or hire one or more sales associates who are good at that process.

We do not cover all aspects of store planning and visual merchandising in this text, but there are some terrific books out there that can be of great help to the new entrepreneur. One of our favorites is *Silent Selling: Best Practices and Effective Strategies in Visual Merchandising*, 3rd edition by Judith Bell and Kate

Ternus (Fairchild, p. 36). In that text, they sum up the importance of planning. They first define the term ***atmospherics*** that describes what the visual merchandiser adds to the store after the carpenters, painters, and electricians have done their work. They then state that "atmospherics should be the result of strategic planning by store owners and managers. Opening and stocking a store is too expensive and competitive to leave any element of store design to chance."

Start with Your Store Size and Limitations

To create the best plan for the new store, the owner must first deal with the basic shape of the store, which can vary greatly from facility to facility. If the store is in a mall, the basic shape might be a typical rectangle with relatively short front and back ends and fairly long sides; this plan is workable but certainly has some limitations. Some of the newer strip malls probably offer a shape that is more square, which provides wide possibilities for product placement but can also be boring if the visuals aren't clever. The real challenges come with the older stores, such as our simulated store on Cary Street in Carytown, Richmond, Virginia (see layouts in Step 2 and on the CD-ROM). These stores were at one time private homes with multiple small rooms and sometimes step-ups to a different room scattered throughout the space. The windows may have been too small when they were homes to be usable for showcasing fashion apparel, and such necessities as customer parking can be an issue. The upside, however, to a space such as ours is that it can be extremely charming and inviting if merchandised in a creative way.

The first challenge when attempting to maximize the appeal of the store shape and selling space is to create a basic traffic plan for the interior of the store. There are a number of basic shapes and variations on those shapes, and it is important to decide during the original store-planning stage about the basic shape of your traffic plan because decisions like where to place carpet with contrasting tile or wood aisles will need to be made during this time. Let's look at some of the more basic traffic patterns that you may decide to adopt for your store:

The grid layout in Figure 9.1 allows for an easy-to-follow floor plan with customer access and the ability to plan the product assortment into four distinctive product departments or classifications, if the owner so desires. When the customers enter the store, they can easily see the four different store areas and the traffic aisles will direct them in to view any one of them, or all of them. In a woman's specialty apparel store, for example, one could easily see the opportunity for one section to display career wear, a second one to highlight sportswear, another for lingerie, and yet another for accessories—or any variation that the product assortment dictates. There is also wide aisle space for the creation of mannequin displays to show off new products, or specialty cases to display ***impulse items*** like jewelry or pretty vases (in a home or gift store). The cash/wrap desk can be positioned either as shown here at the end of one of the aisles or directly in the middle of the aisle formation for good view of all parts of the store (***shrinkage control***) and customer convenience. Focal points for display can be stationed at each end of the four aisles, either with window displays if windows are there or with inside display techniques.

With this basic layout, decisions about floor coverings can be varied but easy. It would make sense to have carpet within the four squares and either wood floors or tile of some sort on the traffic lanes. This kind of plan would further differentiate the departments/classifications of product and would help to naturally guide the customer along the traffic aisles to view everything that the store has to offer.

Limitations to this plan are the lack of flexibility. Suppose a store owner has some categories of goods (heavy winter coats, for example) that need more space in some seasons than others; it would require some imagination to figure out how to keep the departments separate and still house all the product that needs to be displayed during a critical season or time of year. Obviously, this layout works best for a store that is somewhat square in design, although it can be modified to work with some rectangular plans.

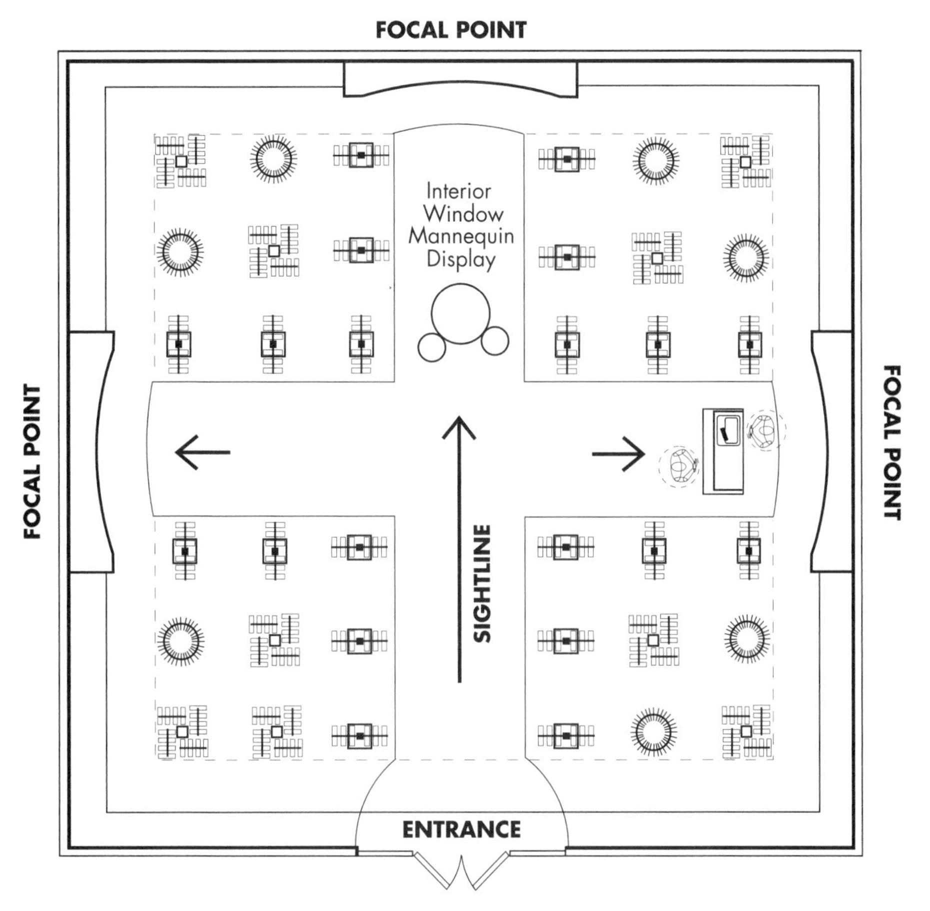

Figure 9.1 *Basic grid floor plan.*

The racetrack floor plan in Figure 9.2 can work with either a square- or a rectangular-shaped store and forms a logical traffic aisle that helps to draw the customer around the whole store. It works for many different types of product and is commonly found in both large- and medium-sized stores. It allows for one or more checkout stations at the front of the store, or they could be housed in the center section of the layout, saving the front area for display of new or particularly trendy merchandise. It also allows for flexibility between seasonal needs and

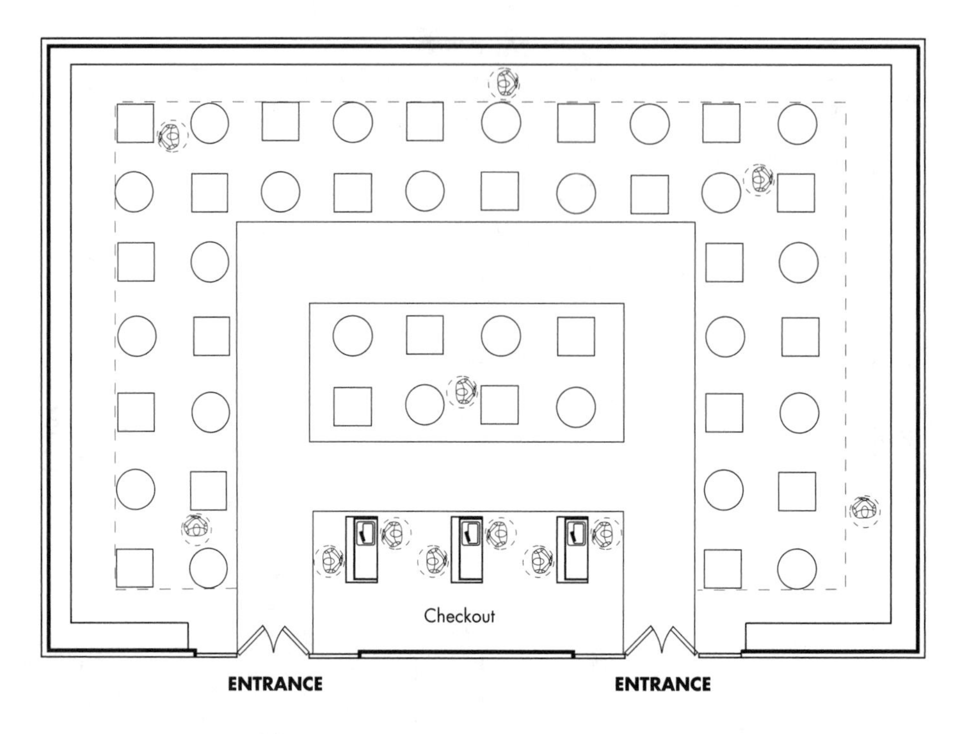

Figure 9.2 *Floor plan layout that features a racetrack traffic aisle.*

shifts in category emphasis. The challenge of this floor plan is to keep it interesting and to avoid having a "discount store look" to your facility. Creative positioning of mannequins and display areas can make all the difference for this plan.

The free-flowing floor plan in Figure 9.3 allows for the greatest store flexibility and is often found in small boutiques. It requires extra-careful planning of fixture placement, especially in the space allowed between fixtures that will encourage shoppers to browse. In the example here, the cash/wrap area is placed in the center of the store, and that is a good location for both customers and staff, but the plan also allows for other good locations for the checkout area if the owner so desires. The traffic aisles, if there are any, would be on the perimeter of the space so the store would conceivably have the same type of flooring throughout.

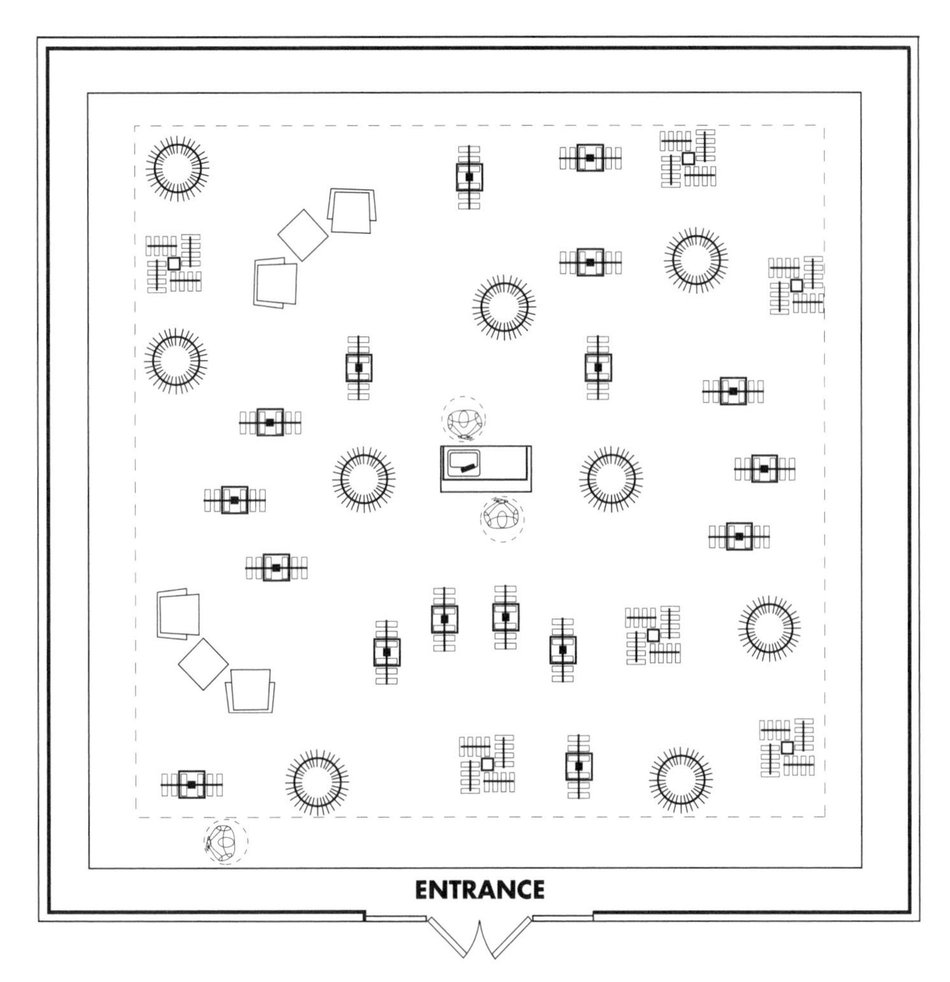

Figure 9.3 *Free-flowing floor layout.*

The soft aisle plan in Figure 9.4 puts emphasis on the wall areas for merchandising and encourages the customer to move easily from one fixture grouping to another. It is also a good plan for grouping categories of product. It allows for mannequin groupings within the selling area and works well for a rectangular store as you might find in a shopping mall. It features wide aisles around the perimeter of the space, and the checkout placement is convenient and does not take away from the selling area.

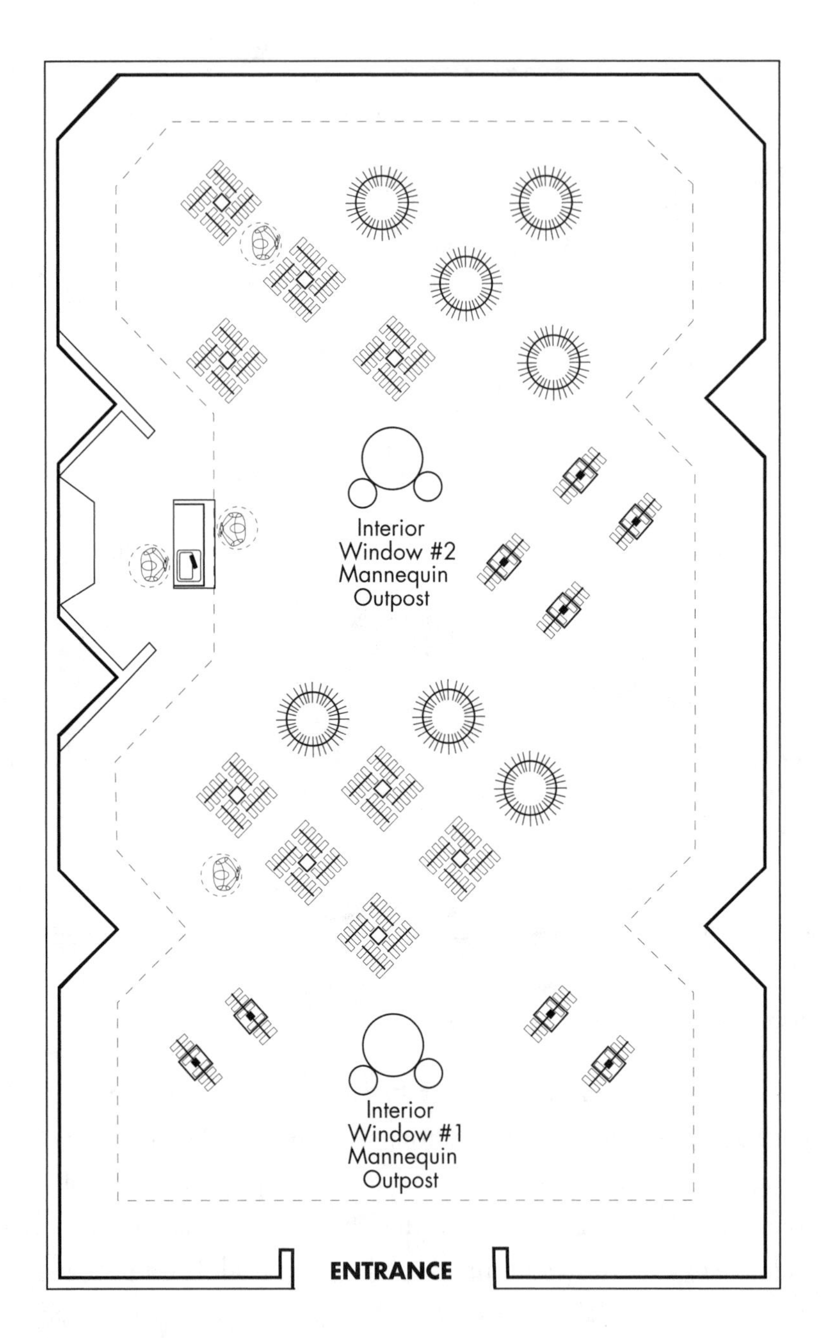

Figure 9.4 *Soft aisle plan.*

Of course, your store might be broken into separate rooms so that a combination of these layouts might work best. In our floor diagram, shown in Figure 9.5, we have two distinctly different areas separated by a wall and steps leading up to the area on the left side of the store. We have chosen a soft aisle plan for our store, which works best to make the best use of our total space.

In addition, we have used our wall space along the sides of the store to the maximum to display our merchandise. We have several built-in display features and a seating area for customer comfort right across from the restroom. The centers of each section in the store are filled with various kinds of fixtures to create interest and good usability for our various kinds of product. The cash/wrap desk is located in a good spot with visibility of the dressing rooms, bathroom, and front door. The problem with this location is that we do not have good visibility of the upper section of the store, but we can access it quickly by the stairs just to the left of the checkout. This plan also allows many interesting spots for mannequin and other types of display. With the help of Jennifer, our consultant, we feel that we have maximized the usage of the space that is available to us. We have kept the back of the store for a storage room and office, and the three dressing rooms are located just outside the stockroom and across from a rack of closeout and discontinued merchandise.

Of course, there is more involved in planning a new store layout than just the traffic aisles. Before starting any construction to remodel the space, the entrepreneur needs to access the wisdom of making upgrades to the areas listed in Table 9.1. Use the table as a guideline when you start your store planning to help you determine what your primary needs are with regard to either cleaning or replacing elements in your new store. Even if you can't replace everything before you open the store, this will help you evaluate priorities and also act as a reminder of other changes that you want to make once your store is open and making a reasonable profit.

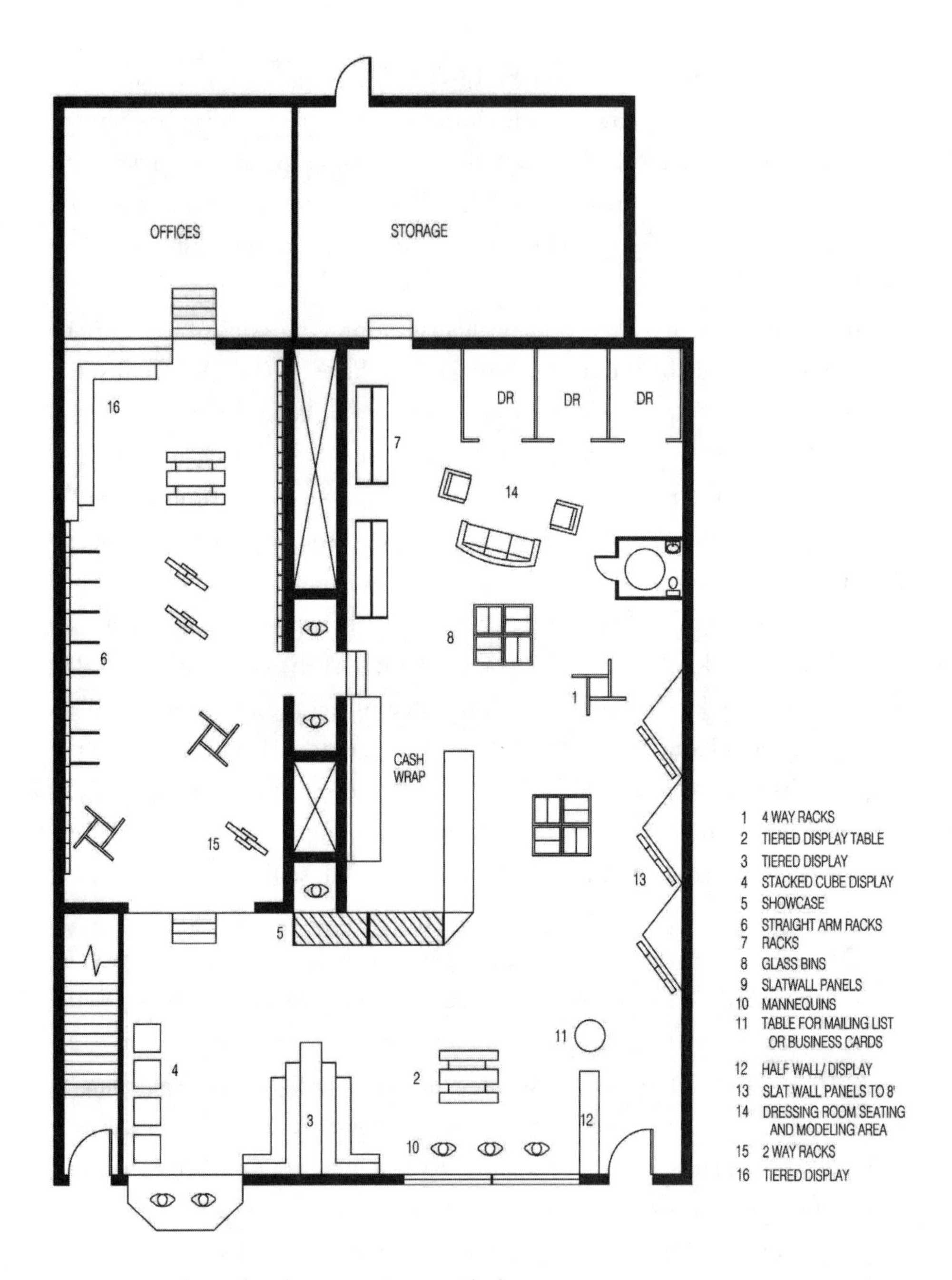

Figure 9.5 *Floor plan for Rose Knows Clothes.*

TABLE 9.1 CHECKLIST FOR STORE REMODEL		
ELEMENT	CLEANING	REPLACEMENT
1. Front of store: windows, sidewalks, paint, awnings		
2. Outside main store sign		
3. Display windows at storefront: Is it big enough? Deep enough? Easy space to create display? Does it need paint?		
4. Floor coverings: Are the current ones usable for your store? Do you want a different covering to better match your image?		
5. Ceiling: Is it too high, too low? Does it need to be covered, have tiles, be replaced?		
6. Lighting: Is it adequate for your needs? Does it set the right mood and match your image? Do you need additional focal point lights?		
7. Cash/wrap area: Is it the right size? Does it have storage areas for gift wrap, supplies, etc.? Is there room to display impulse items?		
8. Dressing rooms: Are they big enough for handicap accessibility? Are they comfortable?		
9. Mirrors: Are they usable? Are they located in the area that matches your store strategy? Do you want them inside or outside the dressing rooms? Are they worn looking?		
10. Traffic patterns: Do the current ones work for your merchandise?		
11. Colors: Do wall, ceiling, and flooring colors marry well with your product? Do they detract from or complement your store image and customer preferences?		
12. HVAC: Have these tested. Do they work properly, or do you need to address the problems with landlord? Do this before you move in merchandise.		
13. Overall cleanliness: Check both inside and outside the store. You don't want to discourage your customer before she even has a chance to shop your merchandise.		

Adapted from information in the Existing Store Checklist in *The Budget Guide to Retail Store Planning & Design* (p. 73) by Jeff Grant (ST Publications).

Working with Store Limitations

A newly built store will likely not present too many limitations inside the store, other than the need to work within the existing space and square footage. You will probably take over a newly built store as a "***vanilla box***" meaning four walls, a ceiling, a floor, a bathroom (unless it is in a mall, in which case the bathrooms might be down the corridor from your store), and maybe a stockroom. You will have to negotiate with the owner of the space for such items as paint in your color scheme, floor covering of your choice, and a customized ceiling with lighting fixtures of your choice. He or she may or may not be willing to work with you on these elements.

An older store may have some real limitations, however. Particularly if your store was once a home or other kind of facility, you will have to decide how you can live with its limitations. You may have too many small rooms, but be unable to remove walls because structurally they are load-bearing walls. You may have a ceiling that is too high for the cozy ambience that you desire. You may have unattractive support beams in the middle of your desired display area, but they are holding up the ceiling or second floor and can't be removed. Like our store, you may have one section of the store that is higher than the others, requiring steps to get up to that space. Like the elevator in the middle of our store, you may have something in the middle of your store that needs to be worked around and worked with (we can't remove the elevator because it is too expensive to remove and it is also one of the access points to the upper floor of our building). These kinds of challenges add to the need to find consulting help, but there are ways to use your store problems to your advantage. For example, that support beam in the middle of the display space can be covered with some attractive material and used to display handbags, hats, or jewelry. In our simulated store, we closed off customer access to the elevators and used the space to house mannequins. There was an ATM at the front of our store; we had that removed, took down the interior walls that closed it off, and built in an attractive display shelf area. Before you make any changes, however, you must check with the landlord and get his or her permission. If the renovations are extensive, you must also get a permit from the city before starting the work.

Stockroom Use

If the store is newly built, the owner might be offered some choice about the size and location of the stockroom if one is included. If not included with the vanilla box, the owner may have to build one. In an existing store, she may or may not have a choice and may have to either live with the stockroom as is or rebuild it to suit the needs of the space. Stockroom size needs will vary from store to store, depending largely upon type of product selection. A shoe store, for example, will need a larger stockroom than other types of stores, as much of the product will likely be kept there and brought out as customer size needs dictate. Because so much of the stock is kept off the selling floor, in most cases, the stockrooms of shoe stores are counted as part of the selling space.

If the stockroom is too large, a new owner might consider using part of the space for an office or for an employee break space. Another option would be to build dressing rooms that take up part of the stock room space. Again, creativity in use of space is called for here.

Planning Inside and Outside to Attract the Target Customer

After the construction is done and the floor layout selected, the new owner should give some thought to store theme. Even if the main aspect of the theme is a color story, it needs to be one that does the best job of showcasing the product and attracting the target customer. Probably everyone knows that children like primary colors, so a children's store will want to make use of those colors in planning store displays, elements such as wall and floor colors, and window displays.

Other themes that the new store may want to incorporate will depend upon target customer and product selections. Here are some ideas from *The Budget Guide to Retail Store Planning and Design* by Jeff Grant:

- Sport stores or sports apparel stores will probably want posters on the walls of athletes; a video showing sports events; bright lighting; vendor banners decorating walls, fixture tops, or hanging from the ceiling; and flexible mannequins that can be posed in active sports positions.
- Lingerie stores will be most appealing with soft-colored walls, floors, and decorating elements to blend with the typically pastel and soft colors of the merchandise. Fixtures may be varied with antique chests and mannequins to give a vintage look, hangers will likely be padded, and Lucite half forms may be used in lieu of or in addition to full-sized mannequins. You may need drawers to house multiple bra and panty sizes and special lighting to create a warm glow.
- Luxury women's apparel stores will also probably use a neutral or soft pastel color scheme to accommodate a variety of clothing for any season. Hangers may be both wooden and padded, depending upon display use. Dressing rooms should be ample to handle lots of clothing items for trying on and also to accommodate all sizes of customers. Mirrors, specialty decorative lighting, beautiful chairs and couches in the sitting area, and elegant fixtures all add to the luxurious feel of the store and the merchandise.
- Shoe store themes can vary widely according to product type and target customer. A specialty sneaker store aimed at a young customer would probably use many of the elements noted above for the active sport store. They, like all shoe stores, however, will want to provide adequate comfortable seating for customer try-ons. A women's shoe store selling designer and better shoes will want a softer, more elegant color story and plenty of mirrors—both floor-type and full-length. The floor should be covered with plush carpet but also have

some other type of flooring in the store where the customer can try the shoes on different surfaces. Fixtures should include risers to showcase individual styles, and window merchandising will be more product-intensive than apparel stores.

- Other types of themes for stores can be Country Antique, Unique Toys, Decade-Specific (a store selling clothes with a sixties feel, for example), Fun at the Pool or Beach for swimwear, Bridal Elegance, and Jewelry—either handmade artisan or luxury items. As you use your imagination, you will see the variations of decor for each type of store.

The important point here is that these decisions need to be made prior to the painting, carpeting, lighting, and ordering of fixtures and other elements that will serve to decorate the store and showcase the merchandise.

Budgeting the Inside and Outside Renovations, Fixturing, and Decor

The message here is short and simple: plan, plan, plan. Don't just run out and start spending money without a clear-cut idea of how much you can and want to spend on renovating and decorating. Don't skimp too much or your customers won't find your store appealing, but don't overspend either. Shop around for the best deals, and don't be afraid to negotiate with fixture suppliers, contractors, and home supply stores. If you are handy with the hammer and saw, or you have friends and family members who can be drafted into service to paint and clean, then you can save a chunk of money. Spend the most money on the renovations that you can't do yourself and on the fixtures and furniture that will be immediately seen and used by the customer. You can save on some of your fixtures and mannequins by buying used ones in good conditions—but you might want to place these in the back of the store until you can afford to buy new ones. You might be able to save by buying used furniture or inexpensive new pieces that require some assembly if these items are to be used in the office or break room. If your store theme accommodates this look, you might be able to refinish or recover old vintage furniture or chests for merchandise display.

Start with the Store Entrance

The impression that your customer has when she sees the outside of your store will be the one that will either bring her into the store or not. The visual impact of that entrance plus her reaction to the look of your store when she first enters will have as much or more to do with your potential to make a sale as the actual merchandise that you offer. If she is turned off before she makes it past the front entrance, your chances are not good that she will even see the merchandise.

The front of the store should be clean and neat and free of chipped paint and broken sidewalk. If there are awnings, window boxes, or other external elements, they too need to be in attractive and working order.

Store Sign

Consider the sign that announces the name of your store. In an existing store, there may a sign that you can reuse, but don't keep it if it is in poor condition or out of character for the image that you want to project. Size is critical: you don't want a sign so small that no one will notice it or see your name, but you certainly don't want a big gaudy sign for a sophisticated clothing boutique. Some locations have restrictions on what kinds and sizes of signs may be used; it is best to check for any restrictions before ordering a new sign. Most important, keep it tasteful.

You can even choose to paint your name and logo on the front display windows, but only if it complements your merchandise assortment. As you can see in Figure 9.6, the storefront of Rumors near the Virginia Commonwealth University campus is eye-catching and gives immediate information about the product within.

Figure 9.6 *The eye-catching storefront of Rumors near the Virginia Commonwealth University campus.*

Display Windows

An assessment of the front display windows is important. Do these windows work for your product selection? This is not usually a problem for stores that were built in recent years for the purpose of retail space; however, converted homes and other buildings might need some construction or reconstruction to make the windows appropriate for the display of fashion product. The display windows for these types of stores serve as the invitation to enter the store and create excitement as they showcase new product. With creative visual merchandising, the windows can create a theatrical way to entice the consumer. They should not be allowed to grow stagnant; it is crucial in a small store with limited selling space to keep the windows fresh and new. Select a theme for the windows; it could be a holiday, the start of a new season, a featured designer, or a new fashion trend. Use appropriate props to display and complement the product.

The question of whether or not to hire a window designer is dependent upon the time, interest, and expertise of the owners and sales associates. Many small store owners make certain that at least one of their full- or part-time sales associates has some talent and experience in this area. If you do not have such a person on staff, you might consider using a window trimming consultant, which you can arrange to pay by the hour or on a monthly contract. You can find ideas and inspiration for store window displays on the Web site www.storewindows.com and for international ideas at www.echochamber.com. Also see the windows of the Pink store on the CD-ROM. The store has won many awards for its windows.

Women's and men's apparel stores that offer better merchandise usually opt for full-size mannequins to display the clothing, as they project sophistication and luxury. There are, however, many other options; the ones that you choose will depend upon the product that you offer and the image that you wish to project. There are half mannequins, body forms without heads or limbs, stick-figure type forms for showing apparel and high-tech pipe forms; Some visual merchandisers just string up the clothing on fishing line flying from the top of the window box or a grid. Use your creativity and do something appropriate but eye-catching. In addition to the mannequins themselves, there is no limit to the variety of props, backdrops, lighting, and other effects that the clever visual merchandiser can include in the window displays. The most important rule to remember is to *change them often* and keep them fresh. Since most of your customers will be loyal, return customers, they deserve some newness and excitement greeting them each time that they come back to shop.

An Invitation to Step Inside and Shop

Assuming that exciting window displays bring the customers into the store, the next thing that they will see are the displays of new merchandise that are set up just to the sides or front of the door. We say "new" because of the importance of keeping this area fresh with new product. The owner would certainly not want to display permanent markdown goods in this prime location unless he was having a store closing event! Not only should the product be new but it also should be rotated often to keep that area interesting. The product on display may not always be something that arrived yesterday, but it can be made to look new by rotating it to a prime location near the door. This is the area that will persuade the customer to move farther into the store and see what else is being offered. For example, in a women's apparel store, the product near the door may not be something that a customer wants to buy, but if it is fashion-right and appealing to the eye, she will be persuaded that there is probably something for her on the racks farther into the store. Just as your traffic pattern will help to guide her along "the yellow brick road," the right placement of appealing merchandise will keep her moving inward and shopping.

Pull the Customer through the Store

As she moves through the store, use creative display to attract her attention and introduce her to new product and ways to put outfits together. This can be done through the following:

- Mannequins or mannequin cluster groupings placed strategically around the store, by the aisles, on the walls, or in a designated alcove. If you don't want to use traditional mannequins, headless T-stands can be used to display outfits.
- Posters and/or backlit photographs of product, such as models wearing outfits formed from product in the store. Posters of sports stars can be used effectively, for example, in a store that sells active sports apparel as a way to create a mood of excitement.
- Fixture topper signs and/or photos that promote a brand or highlight a new trend.
- Use of fixtures themselves to stack lots of versions of an important item. This method is used effectively for items like a new sweater or T-shirt stacked in a glass cube display showing quantities of hot new fashion colors.
- Placing an unusual display piece in the aisle and using it to showcase items such as gift or impulse products.
- Video monitors showing product or setting a mood work well for some types of stores and product lines.
- Don't forget the importance of good lighting to create ambience and show your product to good advantage. No customer wants to try to match colors in a poorly lit store.

Selecting the Right Fixtures

There are basically two types of fixtures that you may choose for your store: custom-made or off-the-shelf ready-made. The decision about which to use depends upon (1) the image that you wish to project, (2) the characteristics of your merchandise, and (3) your fixture budget. Custom-built fixtures are usually the most expensive, although certain types of stores can make use of inexpensive custom-made (homemade) fixtures of materials that one can find in any hardware store. This method would not work for a sophisticated store selling luxury women's wear, but it might be just the thing for a store selling urban sports apparel aimed at a hip young customer. The important thing to remember is to match fixtures to product and image and not to skimp too much if your product commands a higher price point that demands an air of luxury in the store.

We've all been in stores with beautiful custom wood shelving and fixtures, perhaps a store selling expensive men's suits and haberdashery. "For most merchants [however], custom fixtures are economically not only unfeasible but may amount to overkill. More apparel won't be sold because a clothing rack has bleached wood uprights and Lucite display arms rather than standard chrome fittings" (Grant, p. 64).

When making your fixture plan, consider the following important pointers:

- Make sure that your fixtures are flexible and easy to adapt to different seasons, different clothing lengths, and times when the store is fully stocked as well as times when inventory is light.
- Vary fixture types. A store full of round racks or straight one- or two-bar racks looks like a discount haven. All fixtures at the same height create a sea of boredom.
- While variety is good, make sure that fixtures marry together well. They should be of a similar material and style. A combination of pipe racks and expensive wooden fixtures is confusing and looks like the owner bought fixtures at a closeout sale.
- It may make sense to buy some used fixtures, but check them carefully for signs of wear. If they are not up to par, use them at the back of the store to hold clearance or extra stock.
- To decide how many fixtures you will need, design a layout using a computer program or graph paper, or use a consultant to do it for you. Show main aisles with your traffic pattern, and allow enough space between fixtures so that customers can comfortably shop. The space needed will vary: stores catering to young mothers will need space between fixtures for baby strollers; clothing

stores for large-sized customers will need a little extra space between fixtures. A good guideline for fixture spacing is 30 to 36 inches between fixtures and 48 inches wide for major aisles.

- Leave major aisles free. Clutter in the aisles will make it difficult for customers to maneuver through the store.
- In a small store, make use of all space to display product. Consider walls as places to display merchandise; however, be careful not to put product so high that customers can't reach it. If you do put items high on the wall, it is crucial to have enough sales associates on the floor to ensure that they will be available to get them down for the customer.
- If you want to borrow display ideas from a competitor, that's OK, but make sure that your store doesn't end up looking like a "me too." Display and image are part of your niche.
- Typically, smaller fixtures should be placed on the aisles with progressively larger ones behind as you walk toward the back of the store.
- If using gridwall equipment on walls, vary face-out and shoulder-out displays to maximize interest, with shelves placed strategically to hold folded goods.
- Safety first. Fixtures that turn over easily, have sharp edges, or are not securely built can wind up as the paths to a lawsuit.
- "Buy the fixture best designed to present the item to be sold, not just to fill a space on the floor. Remember: Fixtures are silent salespeople" (Lopez, p. 254).

We couldn't possibly cover every type of fixture with the pros, cons, and uses of each, but the following is a recap of the basic varieties of fixtures that apparel stores typically use. You will need to research the types of fixtures that will work best for your store and then compare cost and quality from several sources. There is a lot of information online, and most fixture vendors have a Web site. Before actually making a purchase, it is advisable to visit some various sizes and types of retail stores in your area to observe how well their fixtures would work in your store.

- Freestanding floor fixtures are typically of chrome or chrome and wood or chrome and Lucite. Types that work well for clothing are four-way adjustable-height fixtures with changeable arms and two-way T-stands or waterfall-arm stands. Round racks are used to display quantity, but presentation is limited and the look can imply a discount. Straight rectangular racks with two long hang bars are often used in small stores for items like pants, but wall hanging is probably a more upscale choice. Metal rolling racks can also be used for display but are better used to transport merchandise from the stockroom to the selling floor.
- Slatwall and gridwall systems are ideal for most small apparel stores because they are installed against the walls, take up little space, and are very flexible and changeable. Their panels can also be used to build gondolas and other free-standing floor units if needed. Overuse of these, however, can negatively impact the look of a high-end clothing store if that is the image desired.
- Tempered-glass or wire units with bins or cubes can house quantities of folded sweaters, shirts, pajamas, and other items sold in quantity. Individual display cubes of wood, glass, or other materials can be used to highlight such items as shoes, glassware, or china in a housewares department or gift items in many stores. Grouping cubes together will encourage impulse purchases; use brighter spotlights on them to draw attention (Lopez, p. 257).
- Showcases can perform several useful purposes. Most of these are made of glass, so they complement instead of detract from the look of the product. They are sometimes made with mirrored shelves that give even more of an exclusive look. Often they have locks so that valuable merchandise can be housed in a safe environment. Sometimes they have interior lights.
- Tables of different sizes and shapes can be used for informational cards or literature, to showcase new merchandise or display folded items.
- Built-in towers and shelving units are excellent ways to display goods in quantity, to showcase multiple color choices and organize sizes.
- Specialty fixtures such as tie and belt racks, accessory frames, countertop

jewelry towers, and other varieties can add interest to your selling floor and house difficult-to-display merchandise.
- Vendor fixtures are sometimes available at no cost. These are good for displaying items like bras.
- Furniture and other items can serve as fixtures. They add image and style to the decor of the store, and some merchants also offer these for sale to customers.

Selecting the Right Hangers

The decision about what kind of hangers to buy will depend partly upon whether or not you intend to give them to customers with their purchases; if you do, buy the less expensive ones. If you do not, then you can buy a heavier-weight plastic hanger. You will need to purchase a large quantity with extras, as they will break with frequent use. There are two basic types, and you will need some of each: regular one-bar hangers for dresses, coats, blouses, sweaters, and the like, and hangers with an extra bar across the bottom with clips to hold pants and skirts.

Some vendors will ship garments to you on hangers, but these are usually the cheap plastic type that show up in bargain stores. Almost all boutiques change the merchandise onto their own hangers when the garments arrive in the store. There are also very expensive, custom-type hangers such as the heavy wooden hangers used for men's suits and better women's coats and suits and padded hangers, which are sometimes used for delicate items such as lingerie and fine-gauge sweaters. Some better stores will buy a quantity of these to use in the front of the store or in key areas and fill in the rest of the fixtures with merchandise on plastic hangers. If your image demands these hangers, allow extra money in the budget to buy and replace them. You can find hangers through the same vendors who sell fixtures, either online or through specialty catalogs. Along with hangers, don't forget to research size markers/dividers for the fixtures if you intend to use them.

Other Elements of Interior Display

There are other elements of the store's interior that need to be considered, including color, floor coverings, lighting, checkout areas, seating areas for customers, store signage, office space, and the stockroom.

Color

Select a pleasing color scheme for the store and keep it uniform throughout. Most small stores don't have enough room for two different color schemes for such things as carpet, walls, and trimmings. Pick a scheme that complements your merchandise and projects the image that you wish to convey. If it is a masculine image, you might want to use earth colors, children's stores usually use primary colors, and a sophisticated women's clothing store will usually have a color scheme of neutrals or soft pastels. Keep the seasons in mind; you certainly don't want to change wall and floor colors when the merchandise changes from spring to fall. Think of the color scheme of your store as a backdrop for your merchandise and the product as the most important thing that you want to draw attention to.

Floor Coverings

There are many acceptable choices for floor coverings; they vary greatly in cost, durability, and image projection. If you are taking over an existing location that has an acceptable floor covering, you might elect to clean it and leave it until the store is open and generating a reasonable profit. If the floor is worn or covered in a material that fights your image, you will need to make some changes. This can get expensive. Temporary fixes include polishing up the existing floors and purchasing new or slightly used area rugs.

For stores that are changing the flooring or have a choice for a new store's floors, careful consideration should revolve around such issues as desired location of traffic patterns and what flooring will best add to the ambience and showcase the product. Many small retailers like to use a mix of materials to separate areas of the store or to create traffic aisles. For example, a store selling better women's wear might want wood or ceramic tile or even marble (very expensive) at the entrance and along the main traffic aisles, with carpet on the rest of the floor area.

Lighting

Of course, it is important to have adequate lighting to showcase the product and so that the customer can see and match colors properly. Lighting may also be used to create a mood; if your older store has a beautiful chandelier in the center, you might want to leave it for ambience and style. You can always install track lighting or high hats to add needed light to merchandise areas. Use spotlights strategically for emphasis on display areas and to show off posters, wall photos, and the like. Investments in new lighting are tough choices, as new entrepreneurs might want to hold expenses and elect to just keep the old lighting as is. This may be a fine decision, but don't let the budget keep you from upgrading lighting that will detract from the beauty of your merchandise or create a dark, dingy, undesirable environment.

Checkout Areas

The decision about where to locate the cash/wrap area in a store is largely made to ensure both loss prevention and customer convenience. Often, small retail apparel stores will locate this area near the dressing rooms so that the sales associates can help customers who need different items or sizes. It also allows the sales associates to identify theft.

The size and shape of this area will depend upon type of product. The largest checkout stations are often found in gift and card stores where they use the checkout counters and cases to display impulse items. A women's apparel store often has a checkout station with the cash register in front and another case or table behind where the sales associate can fold the purchases, gift wrap them if desired, and place them in a shopping bag neatly. The area beneath this counter or table will probably hold and hide gift wrap, shopping bags, tissue, and various supplies needed for the cash register. Of course, consideration must be given to how much impulse merchandise is planned for the counter area. Be conservative in this area and do not overcrowd the checkout area with too much product. It will confuse and annoy your customers if they do not have enough room to comfortably perform a payment transaction. If you intend to place small racks near the register to hold impulse purchase items, don't crowd the area. Set them far enough back that the customer doesn't trip over them trying to get to the cash register.

Customer Seating Areas

If the store has the space and the image needs it, a comfortable place for customers to rest while shopping or for companions to sit while awaiting a look at the shopper's choices is a good idea. Certain types of stores almost demand a seating area: bridal salons, for example, almost always have an area for companions to sit while the bride tries on and shows gown choices. Obviously, shoe departments and shoe stores need adequate seating for customer try-on. This can vary, however. For example, owners Rudy Lopez and Billy Manzanares have a small, trendy urban-wear shoe and T-shirt shop named Henry on Broad Street in Richmond, VA. Their customers are young and hip and don't mind sitting on low, covered platforms to try on their shoes. Even if the store does not have room for couches, a few strategically placed chairs would be welcomed by older customers or companions who want to rest while the shopper peruses the racks.

Some stores take the seating area a few steps further with coffee tables filled with current fashion magazines or photos of store product, snacks, and sometimes even a television or video monitor. The monitors can be selling vehicles if they set a mood for the product or show current merchandise on models.

Store Signage

Interior signs are used to designate a particular department, designer brand, or classification, or they can announce a new style or collection. These signs should "have continuity in size, style, placement, and height and must be presented in an attractive manner" (Lopez, p. 259). Owners will also use interior signs to designate sale and clearance items, but be sure to keep those signs attractive and in the same type and image as the rest of the signs in the store; don't make handmade, sloppy signs just because the merchandise is marked down.

Office Space

Here is an area where the owner can conserve on expense. Since the owners and the sales associates will be the ones to use this area, budget furniture or used furniture will work fine, at least in the first years. You will need one or more desks for the owners or managers, perhaps a sitting area with a small table for associate meetings and breaks, some file cabinets, and bookshelves. For the comfort and convenience of the store workers, consider installing a small microwave, coffeepot, watercooler, and refrigerator if there is space for these items. Here, you might also want to locate a fax machine, copy machine, or other office equipment. The owner may also choose to put some of this equipment and break room furniture into the stockroom instead of the office.

Stockroom

In addition to the obvious (merchandise), this room will probably need a garment steamer or iron, adequate shelving to hold product while it is waiting to be prepped for the selling floor, some rolling racks for new apparel, and a few chairs. Other supplies might include a price ticketing machine, if needed; a small sewing machine for repairs or alterations, if they will occur on-site; repair and maintenance tools; and a space for cleaning supplies and equipment.

Your ultimate goal with all visual merchandising is to make your store the customer's favorite place to shop. You want him or her to return often to see "what's new?"—so keep your store and assortments fresh, exciting, and fun.

Finalizing the Store Plan

It is now time to set the plans to paper. You will need a list of all planned expenditures. (You will need these later to put into your Capital Expenditure Plan in Step 10.) You will also need to create a floor plan showing traffic aisles, fixtures, and merchandise placement.

Your budget for the money that you will need to spend to finish and fixture the inside of your store should be divided into these categories:

- Renovations/building
- Fixture and furniture purchases
- Office and stockroom equipment
- Trimming and decorating items
- Cleaning

By categorizing expenses, you will have a better idea of where you might trim expenses if the budget is too large when it is totaled. For our simulated store Rose Knows Clothes, the store renovation and setup budget is shown in Table 9.2. See the Blank Financial Document folder on the CD-ROM and create your own Setup Budget worksheet.

In our example, we listed all the expenses separately and then we tabulated two totals: one for construction and one for fixtures, furniture, office equipment, and trim items. With all of it on paper, we can later see where we might be able to trim expenses if need be. At this point, we have estimated on the high side based on quotes that we received. Once we actually start to make purchases

TABLE 9.2 STORE SETUP BUDGET	
1. Renovations/building	
CATEGORY	**COST**
Build in-store display risers, shelves in stockroom, and dressing rooms	$3,000
Installation of cash wrap, grid-walls, slat-walls, and other fixtures	$1,000
Painting (including cost of paint) in addition to landlord's painting investment & store cleaning	$1,000
TOTAL	**$5,000**

2. Ready-made fixtures and equipment			
ITEM OR CATEGORY	**NUMBER NEEDED**	**COST EACH**	**EXTENDED COST**
Store sign	1	$1,500	$1,500
2-way display racks	3	$40	$120
4-way display racks	3	$80	$240
Tiered tables	2	$400	$800
Stacked cube displays	4	$100	$400
Glass showcases	2	$260	$520
Rectangular straight arm racks	2	$100	$200
Mirror (3-way)	1	$400	$400
Full-length flat mirrors	3	$150	$450
Slatwall system	1	$800	$800
Gridwalls	24 arms and hooks	$2.50	$60
Panels and brackets	30 feet		$350
Shelves and brackets	15	$4.00	$60
Cash wrap and table behind	1	$700	$700
Square glass bins with shelves	2	$300	$600
Mannequins	8	$100	$800
Small table for cards, brochures	1	$100	$100
Couch and 2 chairs—seating area			$600
Countertop display cases	4	(Different sizes & types)	$350
Area rugs for selling floor	3	$1,000	$3,000
TOTAL EXPENSES			**$12,050**

2. Ready-made fixtures and equipment (continued)			
ITEM OR CATEGORY	NUMBER NEEDED	COST EACH	EXTENDED COST
2-person desk	1	$200	$200
File cabinets	2	$125	$250
Chairs	4	$100	$400
Bookcase	2	$100	$200
Small table & chairs	1 set	$200	$200
Microwave	1	$100	$100
Mini refrigerator	1	$100	$100
Coffeepot	1	$50	$50
Scanner/printer	1	$300	$300
Garment steamers	2	$75	$150
Hangers	500	$1	$500
TOTAL			**$2,450**
GRAND TOTAL			**$19,500**

and negotiate work to be done, we will attempt to cut corners where possible (without compromising the look and image of the store). We already know that Jimmie and Rose will do some of the work and will try to get help from spouses, friends, and family to keep down the cost of expenses.

Our next job is to render our floor plan showing traffic aisles, fixtures, checkout station, dressing rooms, bathroom(s), office, storage room, and all other important elements in the store. We will also show mannequin and other display stations. Refer back to Figure 9.5 to see our layout, and please check out our floor plan on the CD-ROM in the Rose Knows Clothes Simulated Store folder. It is certainly not necessary for you to do this for your business plan, but it will give you a better perspective of the way that the floor will look.

Simulation

1. Compile a list of expenses that you will incur in the setup of your selling floor and other areas in the store. Refer to Table 9.2 for an example of one way to do this, but compile your list in a way that works for you and your

plan. Just be careful to list everything that you anticipate spending on the setup of your store. (See CD-ROM for a blank store setup budget form.)

2. Render a plan of your floor layout similar to the one for our store in Figure 9.5. You may use any of the traffic plans shown in this step, or create one of your own. The floor layout should be done on a computer if possible. The folder on the CD-ROM labeled as Illustrator Instructions shows how to create a floor plan using Adobe Illustrator. You should start with the floor plan that you completed as part of the Simulation instructions at the end of Step 2. To this, add the location of fixtures, dressing rooms, stockroom, office, bathrooms, built-in shelving, and all other functional and decorative features of your store interior. Create a legend explaining what each symbol in your plan represents (type of fixture, etc.), and also indicate where your merchandise in each category is housed (four-way fixture holds hanging slacks, for example).
3. Complete Part VI Location in the business plan (see the Introduction) by pulling together the information that you compiled for the Step 2 simulation and combining it with information about the following:
   a. Discuss the benefits of the store layout and visual merchandising that you have planned: type of layout, mannequins, fixtures, and so on.
   b. Describe how your visual merchandising will enhance the saleability of your product selection.
   c. Finalize the prior simulation materials from Step 2 where you addressed such issues as how your location will secure your success and what is desirable about your location.
   d. Include map of store location; floor plan as above (Step 2); completed Table 9.2, Store Setup Budget; completed Table 2.2, Store Location Fact List; and completed lease agreement.

STEP 10
Capital Spending Plan and Cash Flow Forecast

"The only place where success comes before work is in the dictionary." —*Vidal Sassoon*

PROJECT OBJECTIVES

This step in the simulation will guide you through:

- Formulating a capital spending plan for all items that must be purchased and improvements to be made before opening your store.
- Distinguishing the difference between capital items and expense items on the capital spending plan.
- Taking the expenses in the P & L plan and expanding them into a cash flow forecast that reflects the anticipated monthly income and outflow activity for your store.
- Understanding the importance of the cash flow statement as a tool to use when speaking with financial investors and also as a planning and monitoring device to help ensure the health of the business.
- Writing justifications to explain the elements of the capital spending plan and the cash flow statement.

Capital Expenditures List

In Steps 2 and 5, you researched and formulated a fixed-cost expenditures list and transferred this information to the P & L plan. These are the expenses that you must pay every month, no matter how much sales revenue that your store does (or does not) produce. The next task is to put together a list of all the expenses that you will incur before you can open the doors to your new business. These are called capital expenditures, and the total of these will, in large part, determine the amount of the loan that you will need to seek. We think of capital expenditures as "onetime" expenses to distinguish them from fixed expenses, which must be paid monthly. The truth is that you will need to spend money on capital expenditures again over time in your store; for example, computers will become obsolete, fixtures will break and dent, and furniture will need to be replaced as it shows evidence of wear and tear. You will probably not, however, have to replace everything all at once; therefore, your original capital expenditure list is a rare exception in that it will be a long, expensive list and you as store owner must come up with the funds before making even one dollar of sales.

Capital vs. Expense Items

The capital expenditure list is typically divided into two subsections: capital items and expense items. The reason for the differentiation is that capital items are commonly depreciated for tax purposes, and they usually have a life expectancy of more than one year. It is wise to consult your accountant for advice on which expenses belong in this subsection. Entries most commonly found on the capital items list include the following: (McKeever, p. 3)

- Computers, printers, computer software, scanners, and equipment
- Fixtures, racks, office furniture, store sitting-area furniture

- Fax machines, telephones, copy machines, vacuum cleaners, and other small equipment
- Permanent signs
- Alterations and leasehold improvements made to building, bathrooms, walls, storage rooms, and other spaces inside the store

Expense items, on the other hand, represent costs of setting up the business and usually last less than one year. Some of them, however, represent money that you will need to pay as security deposits (rental security and utility deposits, for example), knowing that you will not get that money back until you close, sell, or move the business to another location. Typical entries in the expense items list are as follows: (McKeever, p. 3)

- Opening inventory, most of which you will probably have to pay for either in advance of shipping or COD (cash on delivery at the time of receipt of shipment)
- Business licenses and permits
- Marketing and advertising efforts for preopening promotional activity
- Insurance, and utility deposits and rental security
- Legal and accounting expenses incurred in setting up the business, partnership agreements, and the like
- Office supplies and postage
- Fees to activate utilities, security systems, and to service equipment and clean the store

By the time that the entrepreneur compiles the final list of capital expenditures, she will have completed research, determined the best places to invest her money in these items, and received a firm quote from the dealers. This is not a time to guess on these expenses but to be fairly certain of how much each entry will cost. If uncertain of exact cost, estimate on the high side and round off most of your numbers—no need to list a fax machine as $289.99; $300 is a workable estimate with some wiggle room. While it is important to budget carefully for these expenses, it is also wise to keep in mind the image of the store. As discussed in Step 9 on visual merchandising, the entrepreneur will want to make sure that everything in his store complements the image that he is attempting to convey, and that the target customer will react well to the marriage of product and image. In other words, don't use bargain-basement fixtures, furniture, and equipment inside a store that sells luxury merchandise to elite customers.

When starting the list of capital expenditures, refer back to Step 2 and Table 2.2, Store Location Fact List, to refresh your memory about what kinds of renovations are allowed, what the landlord has agreed to pay for, and what constitutes the tenant's responsibilities. This will give you a good starting place from which to formulate your list.

Capital Expense List Example

For our simulated store Rose Knows Clothes, we started with the floor plan that was developed by our consultant Jen and the expenses that we expect to incur in renovation and furnishing of our store as outlined in Step 9. We tabulated those costs along with all other anticipated expenses that we will have to pay for before our store opens, and we listed them in Table 10.1. (See CD-ROM for a blank version of this table.}

Notice that, since we previously detailed renovation and store fixturing and furnishing costs in Step 9, we listed totals only in this table. At the end of each subsection, Expense Items and Capital Items, we have totaled each, and at the bottom of the form is a grand total for all items in our capital expenditures list. This list will now be transferred to the Preopening column on the cash flow statement.

It is important, at this point, that we explain the items detailed in our

TABLE 10.1 CAPITAL SPENDING PLAN	
EXPENSE ITEMS	**BUDGETED DOLLARS**
Rent/security deposit	$8,500
Electric	$700
Gas and water	$300
Security/Internet	$100
Telephone/fax	$100
Insurance deposit	$500
Supplies/postage	$500
Advertising/marketing	$6,000
Travel	$2,500
Legal advice	$1,000
Accounting/bookkeeping	$500
Inventory	$48,000
TOTAL	**$68,700**
CAPITAL ITEMS	
POS system	$4,000
Telephone/fax	$400
Renovations	$5,000
Fixtures/furniture	$13,000
Store sign	$1,500
TOTAL	**$23,900**
GRAND TOTAL	**$92,600**

capital expenditures list. As mentioned in Step 3, these explanations are called justifications, and they help the reader understand the reasons and rationale for each necessary expense. A look at our justifications in Figure 10.1 will reveal the thoughts behind each entry. The justifications do not need to be long—just the important facts are needed here. Once you have justified an expense, you will not need to justify it again for other financial forms.

Expense Items

- Our lease stipulates one month's rent as a security deposit. In addition, we must pay the first month's rent in advance, before we open the doors.
- Utility expenses include a cost for installation (turning on) and safety check plus the charge for one month's usage of each while we are setting up the store and preparing for opening, and one month's deposit on each.
- The installation of four telephone lines for the store will cost $100, including a line for the fax machine (expense items); the purchase of a cell phone plan with two phones will cost $100 and the fax machine purchase is estimated at $300 (capital items).
- In Step 5, with the development of the first-year P & L statement, we explained the types of insurance that we will need for our simulated store as they were detailed by Jeffrey Virgin of Straus, Itzkowitz & LeCompte. Our plan requires a down payment of approximately $500.
- The budget for supplies and postage includes all the office supplies and cleaning supplies that we will need to buy to get the store up and running.
- Our advertising and marketing budget for preopening is high at $6,000, but we feel that we need to give an extra effort in order to get our target customer into the store. There are many other places for customers to shop in the Greater Richmond area, so we need to get the message out that we are here, we are different, and we want their business. In order to advise all of them at least once that we are in business, we believe that we need to use more expensive vehicles like the *Richmond Times* newspaper, *Style Weekly*, and drive-time radio. We will monitor the success of our opening marketing efforts to make decisions about their value for the future. We have detailed our planned expenditures in Section VII on marketing.

Figure 10.1 *Capital Spending plan justifications.*

- Both partners will need to make a trip prior to opening to the MAGIC show in Las Vegas in order to purchase inventory and to do fashion research. We estimate this cost to be $2,500.

- To set up our accounting records, review our lease, and set up our business arrangement, we will need the services of our accountant and lawyer. We estimate those costs to be $1,500.

- The cost to purchase our inventory will be about $48,000 (refer back to Step 6, Table 6.2). This is an expense item because the theory is that we will sell out of our opening inventory within the first year of business. Of course, we can never let our store completely sell out of inventory, so we will plan and make additional inventory purchases throughout the year. These will not be shown in the start-up column; they will be shown in the months in which we plan to pay for the inventory shipments.

- This brings our expense item totals to $68,700.

Capital Items

- Two cordless telephones for the store will cost $100; a fax machine will cost $300.

- Point-of-sale system from Isis, Inc. in Richmond, Virginia, costs $4,000. This includes $3,000 for hardware, including cash drawer, barcode scanners and equipment, computer terminal, check scanner and credit card reader, inventory system, accounts receivable and payable systems, plus $1,000 for software to support the system.

- We have allotted $1,500 for the purchase of a sign. We will need a new sign to replace the one that was on the store prior to our lease.

(continued)

- Renovations for our store will primarily be paid for by the owner of the store as he turns the building into two separate rental store spaces (discussed in Step 2). In addition to the major renovations (for which we will pay a higher monthly rent and that is merely a simulation for this textbook), we will need to do some additional renovations to the space. These have been detailed in Step 9 on visual merchandising, but primarily involve building shelves in the storage room, installing built-in display units in the store, and building display shelves in the store. As stated in Step 2, we elected to spend $1,000 to hire a consultant to help us plan the layout for our store. We helped to minimize the additional costs for installation and building by doing some of the work ourselves and having friends and family assist. We did hire a carpenter to do some of the work that we could not do. Total renovation costs for us: $5,000.
- Fixture and furniture costs were detailed in Step 9 and will cost us $13,000 in total for store fixtures, furniture, office appliances, and equipment.
- The total amount allotted for capital items is $23,900, bringing our total for the capital spending plan to $92,600.

Expense Control

- Each partner owns a laptop computer, and we will use these for the business until we are able to afford to buy PCs that are business-owned.
- At least initially, each partner will use his or her personal cars for the business.
- The partners and store associates will take turns cleaning the store, at least in the first year. After our store has proved to be successful, we may elect to hire a cleaning service.

Cash Flow Forecast

Cash flow is the amount of money that flows in and out of a business during a particular accounting period. For this simulation, we will be working with the amount of cash that comes into the business monthly and the amount that must be paid out monthly for fixed costs, to purchase inventory, and to pay back loans. Each cash flow statement represents one year and can be stated as either a calendar or a fiscal document. In the case of our simulated store, we are operating on a fiscal-year basis, starting with the first month that our store opens and continuing one year from that month. Since the cash flow statement follows the P & L statement and works from the information formulated there, both documents should use the same financial planning calendar.

The cash flow statement is the most important financial document in the entrepreneur's business plan, as it will determine not just whether or not he will produce a profit at year end, but more important, it will answer the question, "Will he be able to pay his bills each month?" While we never want to diminish the importance of producing profit, it is more important in the first year especially that the owner be able to pay all fixed costs, purchase additional inventory as needed, and pay back start-up loans on time. In other words, there is enough money coming in each month to take care of all the money that must go out of the business in the same time period.

One important fact to understand about the cash flow statement is that it reflects activity for the month; incoming funds are noted as they are expected to be received into the business, and outgoing funds are shown for the month in which bills will be paid, not necessarily the month in which the debt is made. For inventory especially, you may be able to arrange terms with your suppliers in which you can pay for the merchandise 30 days after it was shipped from the supplier. The merchandise orders may be placed in January and shipped and received in February, but you might not have to pay for them until March (in reality, the time between orders and delivery is usually much longer, as so many products in today's market are made in offshore factories). On your cash flow statement, the entry for that inventory would fall under the month of March, the month in which you will be required to pay for it.

Cash Flow Forecast Example

Take a look at the cash flow forecast for our simulated store Rose Knows Clothes, shown in Table 10.2. The cash flow forecast (also called statement) is divided into two sections: cash-in and cash-out.

Cash-In

Cash–in is a combination of receipts that represent beginning cash (same as ending cash from the month before) plus planned sales revenue for the month plus any injections of cash from loans or lines of credit. These are totaled on the line marked "Total Cash."

Cash-Out

Cash-out is made up of fixed costs that must be paid during each particular month plus loan repayments and payments on the ***line of credit***, if applicable. Cash-out expenses are totaled on the line marked "Ttl Exp," which is an abbreviation for "total expenses." "***Below the line***" (an accounting term meaning expenses that are contingent and are shown below the Total line) is a line for ***Owner's draw***. This is the place to show the salary/payments that are planned to be paid to the owners of the business each month. The reason for showing them below the line is that this is the first place that an owner will cut expenses if sales do not materialize as planned. Entrepreneurs should be prepared for this possibility when they decide to open a business. You will not be able easily to cut fixed expenses, but you can certainly take less money for yourself in a particular month; hopefully, you can make it up in a more profitable month.

After the owner's draw is added into total expenses, a new total called "Cash Payments" is calculated. That number is subtracted from the Total Cash line above, and an "EOM (end-of-month) cash" number is determined. The EOM cash number at the end of one month (April, for example) becomes the cash number on the first line of the next month (May, for example) and is listed next to "Cash" for the applicable month.

Preparation

When attempting to complete the cash flow forecast, one needs to pull together for reference: the three years of P & L statements previously prepared and the capital expenditures list with details of all purchases and expenses that the entrepreneur will need to pay prior to opening.

After setting up the forecast with columns for Preopening (Start-up) Expenses and then one for each month of the first year fiscal calendar followed by a Total column, enter the monthly planned sales revenue on the appropriate line under each of the months in the plan. Then enter the total first year's planned sales revenue under the Total box to the far right on the appropriate line.

Start with the Preopening Expenses

Begin with the column marked "Preopening" and enter first, on the line at the top marked "Cash," the amount of money that the new owner(s) are prepared to invest into the business from their own funds. It is important that the cash amount indicated here be easily explained as to the origin of that money. If the personal financial statements of the owner(s) reflect enough cash and other liquid assets to justify this investment without "breaking the bank," then no justification is required. If the money is to be a gift from a friend, spouse, or relative, you must divulge that information in the justifications to the cash flow statement. If the money is to come from a private loan, the terms of that loan and a copy of any promissory note must be produced. Keep in mind that

TABLE 10.2 CASH FLOW FORECAST

	PREOPENING	APRIL	MAY	JUNE	JULY	AUGUST
Cash	$30,000	$7,400	$16,096	$18,283	$18,655	$18,507
Loan	$100,000					
Line of credit	$30,000					
Gross sales		$23,100	$30,690	$27,390	$23,100	$27,390
TOTAL CASH	**$130,000**	**$30,500**	**$46,786**	**$45,673**	**$41,755**	**$45,897**
ITEMIZED EXP.						
Staffing		$2,100	$2,450	$2,200	$1,700	$2,200
Payroll taxes		$294	$343	$308	$238	$308
Rent	$8,500	$4,250	$4,250	$4,250	$4,250	$4,250
Electric	$700	$330	$330	$330	$330	$330
Htg & water	$300	$105	$105	$105	$105	$105
Maintenance		$100	$100	$100	$100	$100
Sec/Internet	$100	$100	$100	$100	$100	$100
Phone/fax	$500	$175	$175	$175	$175	$175
Banking		$200	$200	$200	$200	$200
POS system	$4,000	$100	$100	$100	$100	$100
Insurance	$500	$260	$260	$260	$260	$260
Supp/post	$500	$200	$200	$200	$200	$200
Marketing	$6,000	$1,200	$1,200	$1,000	$800	$900
Travel	$2,500	$100	$100	$100	$100	$2,000
Acct/bkp	$500	$600	$300	$300	$300	$300
Misc		$200	$200	$200	$200	$200
Inventory	$48,000	$-	$14,000	$13,000	$10,000	$15,000
Legal	$1,000					
Renovations	$5,000					
Fixt/furniture	$13,000					
Store sign	$1,500					
Loan repay		$1,090	$1,090	$1,090	$1,090	$1,090
LOC repay						
Ttl exp	$92,600	$11,404	$25,503	$24,018	$20,248	$27,818
Owner's draw		$3,000	$3,000	$3,000	$3,000	$3,000
TOT CASH PMTS		**$14,404**	**$28,503**	**$27,018**	**$23,248**	**$30,818**
EOM CASH	**$7,400**	**$16,096**	**$18,283**	**$18,655**	**$18,507**	**$15,079**

	SEPT	OCT	NOV	DEC	JAN	FEB	MARCH	TOTAL
	$15,079	$17,472	$16,744	$13,434	$10,678	$8,608	$2,620	
	$32,010	$27,390	$31,350	$36,300	$20,790	$23,100	$27,390	$330,000
	$47,089	**$44,862**	**$48,094**	**$49,734**	**$31,468**	**$31,708**	**$30,010**	
	$2,550	$2,200	$2,500	$2,900	$1,500	$1,700	$2,200	$26,200
	$357	$308	$350	$406	$210	$238	$308	$3,668
	$4,250	$4,250	$4,250	$4,250	$4,250	$4,250	$4,250	$51,000
	$330	$330	$330	$330	$330	$330	$330	$3,960
	$105	$105	$105	$105	$105	$105	$105	$1,260
	$100	$100	$100	$100	$100	$100	$100	$1,200
	$100	$100	$100	$100	$100	$100	$100	$1,200
	$175	$175	$175	$175	$175	$175	$175	$2,100
	$200	$200	$200	$200	$200	$200	$200	$2,400
	$100	$100	$100	$100	$100	$100	$100	$1,200
	$260	$260	$260	$-	$-	$-	$-	$2,080
	$200	$200	$200	$200	$200	$200	$200	$2,400
	$1,200	$1,100	$1,300	$1,500	$900	$1,000	$1,100	$13,200
	$100	$100	$100	$100	$100	$2,000	$100	$5,000
	$300	$300	$300	$300	$300	$300	$600	$4,200
	$200	$200	$200	$200	$200	$200	$200	$2,400
	$15,000	$14,000	$20,000	$24,000	$10,000	$14,000	$15,000	$164,000
	$1,090	$1,090	$1,090	$1,090	$1,090	$1,090	$1,090	$13,080
	$26,617	$25,118	$31,660	$36,056	$19,860	$26,088	$26,158	$287,468
	$3,000	$3,000	$3,000	$3,000	$3,000	$3,000	$2,000	$35,000
	$29,617	**$28,118**	**$34,660**	**$39,056**	**$22,860**	**$29,088**	**$28,158**	
	$17,472	**$16,744**	**$13,434**	**$10,678**	**$8,608**	**$2,620**	**$1,852**	

lenders will be more willing to loan money to entrepreneurs who are willing to risk their own assets on the business. If an entrepreneur has no assets and must borrow money even for the down payment on the business, then he is at a disadvantage when asking for a loan of more money to start the business.

Working down the Preopening column, skip the lines marked "Loan," "Line of credit," and "Total Cash." You will fill those in later. Next, after the heading marked "Itemized Exp" (abbreviation for "expenses"), enter the expenses listed in your capital expenditure plan. You may combine some of the expenses that you listed separately if you wish, but only those that fall under "capital items." Expense items that also have corresponding "fixed expenses" entries on the P & L should be listed here separately (but also to reflect the way that they were listed on the P & L statement). At the bottom of this list, total all capital expenditures and enter that total on the line labeled "Ttl Exp" (abbreviation for "total expenses").

Now go back to the top of the column and subtract the amount of money that the owners plan to put into the business from the total that you just calculated for Ttl Exp. It is at this point that a conversation with your banker, SBA official, and/or financial backers should take place, as some degree of judgment and evaluation is needed to determine the requested amount of your loan. Let's look at an example.

Calculating the Loan Amount

On our simulated cash flow forecast for Rose Knows Clothes, the owners have decided to put $30,000 between them into the business. At this point, it should be noted that the contribution from each partner does not have to be the same. In the legal contract that is drawn up between the new partners, however, stipulations should be detailed concerning the financial contributions from each partner and the resulting degree of ownership—or the terms for making the financial investments equitable. The following three options exist:

1. One owner can be legally designated to have a greater share of the business. If Rose can only contribute $10,000 and Jimmie must pay $20,000 of the investment to start the business, it may be determined that Jimmie owns a greater share of the business (and resulting liabilities) than Rose, and Rose may legally be a lesser partner.
2. An agreement can be made that Jimmie will realize a higher owner's draw than Rose until such time as her investment equals that of his.
3. The extra $5,000 that Jimmie puts into the business can be contracted as a loan to Rose; terms can be arranged for the payback at a specified interest rate.

In our simulated plan, our Ttl Exp line indicates a need for $92,600 to pay the expenses of opening our store. If we contribute $30,000 to the business start-up, that would leave us with a shortfall of $62,600 to pay these necessary expenses. We could certainly request a loan of $65,000 to cover everything, but that would leave us with very little extra money with which to open and to cover the contingencies of such things as a shortfall in planned sales in the early months. Here, as in other stages of the business plan development, critical thinking and decision making must be employed. We must deal with two important truths that seem contradictory: (1) the bigger the loan, the higher the monthly payback installment and the more months that we will potentially be indebted to the financial community versus (2) if we have no extra cash in the business from the beginning, we might fail before we even have a chance to start. The important word here is "balance," and the rationale behind the calculation of the desired loan amount must be explained in the justifications to the cash flow statement and in other parts of the business plan (notably, the beginning summary or letter of intent—to be covered in Step 12).

At the very bottom of the cash flow forecast is a space for EOM (end-of-month) cash. This is the money that remains after everything else in the Preopening (Start-up) column is addressed. It is the result after the line marked

"Ttl Exp" (in our case, $92,600) is subtracted from the line near the top of the column marked "Total Cash." The Total Cash line is the result of the addition of cash (our investment in the business) plus the amount of the loan (yet to be determined), plus the amount of the line of credit (yet to be determined). That very last dollar figure next to EOM cash will also be the first dollar figure at the top of the column for our first month of opening (in our case, April). In other words, after we pay for all our opening expenses (including paying for our start-up inventory), this will be the money that is left from all the money that has been poured into the business before opening (Our cash + Bank's loan + Any line of credit that we are able to secure). When we look at that dollar figure next to EOM cash at the bottom of the Preopening column, we want to be sure that we have some cash to start the next month, but not so much cash that we will force ourselves into too high a loan amount. Confusing, isn't it? That's why we advise that you seek the counsel of your team of experts: banker, accountant, lawyer. As a rule of thumb for students, that final cash amount left after all bills are paid should (for most small start-up businesses) not exceed $10,000.

Keep in mind that you would never request a bank loan for an uneven amount. For example, we would not go into a bank and ask for a $62,600 loan to start up our store. We would round off that request to an even dollar amount. You will also not be very successful if you request a loan for a small business that is greater than your anticipated sales revenue for the first year. Be reasonable; don't expect the investors to take that large a risk, and don't put yourself into a position that it will take so long to pay back the loan that you will be living on tuna sandwiches for years.

Completing the Columns

When you complete the balance of the columns for each month on the cash flow statement, it is very important to note the "Total" column on the far right of the form. These totals are the addition of all entries in a particular row for the months only. The totals do not include anything in the Preopening column.

Do not add those numbers into any entries going across the form; just add those numbers down the column to be able to calculate your loan amount.

Will You Need a Line of Credit?

An operating line of credit (often abbreviated as LOC) is a reserve of money that a bank or lending institution has agreed to hold for you until a time (usually an emergency) occurs when you will need to borrow from that credit pool. In a way, it is like a checking account held aside for you until you need it. The only difference is that once you begin to draw against those funds, you must either start to pay those funds back each month or, at least, pay interest monthly on the money drawn. Also, your "draws" may not exceed the total amount in the line of credit reserve; if you need more, you must go back into negotiations with the bank. If you never use the funds, however, you will not need to make any payments to that credit line. Most credit line reserves do have rules about frequency of usage, cancellation if not used by a certain date, and other stipulations. Make sure that you understand all the rules before you sign to take the line of credit reserve. Most new businesses will need a line of credit for the first few years until sales revenue begins to exceed fixed expenses by a wide margin.

There is another, newer way to take advantage of a line of credit, and that is to use a portion of the initial loan like a line of credit. In other words, the line of credit use and payback will remain the same, but we will hold (in our example), a portion of the initial loan as a line of credit, instead of applying for the LOC as a separate loan. We will only draw from that LOC if we need it, and only make payments against it if we use it. Otherwise, the money will stay with the bank. In our example cash flow forecast, we project that we will not need to draw from the LOC for our first year of business. However, if we need it, it is available to us immediately. This is a real comfort and helps the new owner to sleep better at night. At the end of this step is an example of our cash flow statement as it might look if we did not make our projected sales in the first few months of business, so that we might need to draw from the LOC.

Finalizing the Amount of the Loan

Using all that we have just discussed, let's look again at the loan amount. In our simulated example, we wanted to make sure that we had some cash left at the end of the Preopening column (we figured that anywhere between a few thousand to just under $10,000 would work). We wanted to hold some of the original loan amount in the form of an LOC (see line 3 at top of form). We didn't have a set amount in mind; however, we thought that it should be at least $10,000, but certainly not more than $50,000. We settled on $30,000 because it seemed reasonable, after looking at our $62,600 shortfall in paying start-up expenses and adding in a few thousand dollars for EOM cash, that a total loan of $100,000 would work for our needs. That gave us the $62,600 shortfall that we need plus a $30,000 reserved LOC amount for the bank to hold for emergencies, and still left us with $7,400 in EOM cash to start the next month.

As you can see, there is no exact formula for assessing the loan amount needed. It is a matter of reason and common sense. We also feel that a $100,000 loan is reasonable for the size of our business, that it is an amount that we can pay back from sales revenues (if all goes as planned) in a few years, and that the monthly payback will not jeopardize our ability to realize monthly cash flow or cause us to have to eat every night at Mom's house.

Calculating Loan Payments

Now that we have established a loan amount, how will we pay it back?

Referring back to Step 3 for ways to fund our business gave us a start toward finding funding. We have decided to work with a local bank in Richmond; they have quoted us a 10.5 percent interest rate (2 percent over ***prime***) for a period of eight years to pay it back at an amortized monthly payment (principal + interest) of $1,090 per month. We have indicated the monthly payment on the line on the forecast marked "Loan Repay."

Should we have to take money from the LOC, we will indicate the amount taken in the month that we will use the money on the second line from the top of the Cash Flow Forecast marked "Loan." In the first month that we take the money, we will pay nothing against the loan amount. In the next month after taking money from the LOC, however, we will have to either pay some of the money back plus interest from the month before or we will at least have to pay interest even if we don't pay any against the principal amount that we owe on the LOC. Table 10.3 shows what would happen if our sales forecast for the first few months of opening was lower than anticipated, resulting in the need to take money from the LOC.

What Happens to the Cash Flow If You Fall Short of Sales Projections?

This revised cash flow statement shows what can happen if we miss our sales projections for the first two months of business. In this example, we have shown a rather significant shortfall for April and May and have left sales projections for all other months the same as our original plan. This is meant merely to show what can happen if we have to use our LOC.

Notice that we would have had a loss in May for EOM Cash if we had not taken the $1,000 from the LOC (shown on line three under May). In a business, this cannot happen because we still need to pay our bills each month. Yes, we could have reduced the amount of our Owner's Draw in order to make up the shortfall, but we want to show what can happen if one draws from the LOC. Observe that we have shown no indication that we must pay anything against the $1,000 in the month of May, but we must pay interest on that LOC amount in the next month—June. We calculate the amount of interest by taking our annual interest rate (10.5 percent) and dividing by 12 months, resulting in a monthly interest rate of .009 percent. That tells us that we must pay $9 against the loan in the month of June. We continue to pay this interest amount each month until August, when we had to take another draw of $4,000; in September our EOM cash line shows that we will have some extra EOM cash, so we decide to pay $2,000 against the principle amount, but we still must pay the

TABLE 10.3 REVISED CASH FLOW FORECAST						
	PREOPENING	**APRIL**	**MAY**	**JUNE**	**JULY**	**AUGUST**
Cash	$30,000	$7,400	$7,996	$493	$856	$699
Loan	$100,000					
Line of credit	$30,000		$1,000			$4,000
Gross sales		$15,000	$20,000	$27,390	$23,100	$27,390
TOTAL CASH	**$130,000**	**$22,400**	**$28,996**	**$27,883**	**$23,956**	**$32,089**
ITEMIZED EXP.						
Staffing		$2,100	$2,450	$2,200	$1,700	$2,200
Payroll taxes		$294	$343	$308	$238	$308
Rent	$8,500	$4,250	$4,250	$4,250	$4,250	$4,250
Electric	$700	$330	$330	$330	$330	$330
Htg & water	$300	$105	$105	$105	$105	$105
Maintenance		$100	$100	$100	$100	$100
Sec/Internet	$100	$100	$100	$100	$100	$100
Phone/fax	$500	$175	$175	$175	$175	$175
Banking		$200	$200	$200	$200	$200
POS system	$4,000	$100	$100	$100	$100	$100
Insurance	$500	$260	$260	$260	$260	$260
Supp/post	$500	$200	$200	$200	$200	$200
Marketing	$6,000	$1,200	$1,200	$1,000	$800	$900
Travel	$2,500	$100	$100	$100	$100	$2,000
Acct/bkp	$500	$600	$300	$300	$300	$300
Misc		$200	$200	$200	$200	$200
Inventory	$48,000	$-	$14,000	$13,000	$10,000	$15,000
Legal	$1,000					
Renovations	$5,000					
Fixt/furniture	$13,000					
Store sign	$1,500					
Loan repay		$1,090	$1,090	$1,090	$1,090	$1,090
LOC repay				$9	$9	$9
Ttl exp	$92,600	$11,404	$25,503	$24,027	$20,257	$27,827
Owner's draw		$3,000	$3,000	$3,000	$3,000	$3,000
TOT CASH PMTS		**$14,404**	**$28,503**	**$27,027**	**$23,257**	**$30,827**
EOM CASH	**$7,400**	**$7,996**	**$493**	**$856**	**$699**	**$1,262**

	SEPT	OCT	NOV	DEC	JAN	FEB	MARCH	TOTAL
	$1,262	$1,610	$855	$3,518	$1,681	$530	$1,461	
			$6,000			$6,000		$15,000
	$32,010	$27,390	$31,350	$36,300	$20,790	$23,100	$27,390	$311,210
	$33,272	**$29,000**	**$38,205**	**$39,818**	**$22,471**	**$29,630**	**$28,851**	**$356,571**
	$2,550	$2,200	$2,500	$2,900	$1,500	$1,700	$2,200	$26,200
	$357	$308	$350	$406	$210	$238	$308	$3,668
	$4,250	$4,250	$4,250	$4,250	$4,250	$4,250	$4,250	$51,000
	$330	$330	$330	$330	$330	$330	$330	$3,960
	$105	$105	$105	$105	$105	$105	$105	$1,260
	$100	$100	$100	$100	$100	$100	$100	$1,200
	$100	$100	$100	$100	$100	$100	$100	$1,200
	$175	$175	$175	$175	$175	$175	$175	$2,100
	$200	$200	$200	$200	$200	$200	$200	$2,400
	$100	$100	$100	$100	$100	$100	$100	$1,200
	$260	$260	$260	$-	$-	$-	$-	$2,080
	$200	$200	$200	$200	$200	$200	$200	$2,400
	$1,200	$1,100	$1,300	$1,500	$900	$1,000	$1,100	$13,200
	$100	$100	$100	$100	$100	$2,000	$100	$5,000
	$300	$300	$300	$300	$300	$300	$600	$4,200
	$200	$200	$200	$200	$200	$200	$200	$2,400
	$15,000	$14,000	$20,000	$24,000	$10,000	$14,000	$15,000	$164,000
	$1,090	$1,090	$1,090	$1,090	$1,090	$1,090	$1,090	$13,080
	$2,045	$27	$27	$81	$81	$81	$135	
	$28,662	$25,145	$31,687	$36,137	$19,941	$26,169	$26,293	$287,468
	$3,000	$3,000	$3,000	$2,000	$2,000	$2,000	$2,000	$32,000
	$31,662	**$28,145**	**$34,687**	**$38,137**	**$21,941**	**$28,169**	**$28,293**	
	$1,610	**$855**	**$3,518**	**$1,681**	**$530**	**$1,461**	**$558**	

interest on the amount we owed at the end of August ($5,000). Before the fiscal year is over, we have accumulated $15,000 in money owed against the LOC (see the Total column on the far right of the line for Line of Credit). Note that we have also reduced the amount of the owner's draw from $3,000 to $2,000 for the last four months of the fiscal year, and we still only show an EOM cash amount at the end of March of $558.

This demonstrates what can happen if a new store misses its sales estimates for a couple of months. In reality, if this happened, we would probably take the following actions:

1. Reduce the amount of the owner's draw earlier in the fiscal year.
2. Reevaluate our inventory purchases and try to either delay some deliveries or cancel future orders to reflect the lower sales trend.
3. If things do not improve soon, we might attempt to renegotiate the terms of the original loan to allow us more years to pay it off, and thus reduce our monthly payments.

Planning Inventory Purchases

Let's discuss the purchase of inventory. In Step 6, we developed a plan for opening inventory purchases. We also talked about the fact that, after the store has been open for a few months and you establish some baselines for purchases, you would begin to use the six-month buying plan to purchase your future inventory. This cash flow forecast shows projections of the way that inventory purchases would flow. In reality, you would not place orders all at once for a whole year's worth of inventory. The owner would buy opening inventory and then place future orders for about a month or two out—until she could determine how well the product will sell and move out of her store. When planning purchases for the rest of the fiscal year (before making a six-month buying plan), the entrepreneur should base purchase amounts on the COGS figures from the

P & L statement. These monthly COGS numbers, after all, represent the cost value of the inventory that you are selling in a particular month. On our simulated P&L statement for year one, for example, we are estimating that we will move $169,289 worth of stock at cost ($11,250 of this amount within the first month of opening in April). Since our original opening inventory was $48,000, then we should plan to receive more inventory in the month of April (we decided to receive about $14,000 more because our May COGS estimate is $14,946 and May is a month with a higher sales estimate than April). Please note that we have entered that $14,000 inventory purchase not under the month of May but under the month of June, since that is when we will pay for the goods. Each succeeding month can be handled in the same manner, comparing COGS estimates on the P & L plan with inventory already in stock. At the end of the fiscal year, the total of all our purchases for the year (excluding the inventory in the Preopening column) should be close (not exact) to the annual total of our COGS from that year's P & L plan. It should be mentioned that inventory does not have to be received every month; it would depend upon your product lines, but most small apparel boutiques would choose to receive goods in this manner—to improve cash flow and turnover, and because of small stockrooms.

Simulation

1. Prepare a capital spending plan similar to Table 10.1. (See your CD-ROM for a blank version.)
2. Write justifications for the plan similar to those in Figure 10.1.
3. Prepare year one of the cash flow forecast (see Table 10.2). (See your CD-ROM for a blank version.)

4. Complete years 2 and 3 of the profit and loss statement.
5. Complete years 2 and 3 of the cash flow forecast.

STEP 11
Financial Documents for Evaluation and Tax

"I'm an optimist, but an optimist who takes his raincoat."

—Harold Wilson

This step was written by Arthur E. Collins, Jr., Director for Government Contracting, United States Small Business Administration, retired.

PROJECT OBJECTIVES

This step in the simulation will guide you through:

- Understanding the necessity of ***pro forma*** financial statements
- Steps to completing the ***income statement***
- Using the income statement
- How to prepare a balance sheet for the new business
- Evaluating the uses and needs for a balance sheet
- Understanding and using key ratios to measure success

Preparation

Throughout this simulation, we have examined many facets of planning for a new business. To do this, we have presented a number of planning concepts and worked through an equal number of planning documents. Some of the most critical exercises have involved making financial projections. A good example is a cash flow forecast. Because we are simulating a business launch, our planning documents have been entirely future-focused. Now we are at a point where we must turn to summarizing our plans in formats that are commonly recognized and readily understood. The summaries that will draw together the key elements of our plans and communicate the essence of what we are trying to accomplish are called ***pro forma financial statements***.

Understanding and Using Pro Forma Statements

What are pro formas? Very simply, they are balance sheets, income statements, and cash flow statements that, when taken together, are a budget or a *hypothetical* financial model of a business. Pro forma financial statements can be prepared for existing businesses or for new businesses. If pro formas are prepared for an existing business, they are generally intended to assess the effects of some sort of change, perhaps in the business model or in the environment.

In the case of an existing business, the hypothetical model pertains to a future period, given a certain set of assumptions. These assumptions may range from continuity of existing conditions to a radical alteration of conditions, such as entry of the business into a new market. An example of the former would be continued operation of a four-store midpriced ladies' casual clothing chain in the metropolitan Richmond market over the next year, with stores in Chesterfield Town Center, Regency Square, Virginia Center, Commons and Short Pump Town Center. An example of the latter would be expansion of the same chain into a fifth market—Carytown—at a higher price point, in a free-standing storefront, specializing in high-fashion women's jeans.

Over time, historical performance, as documented in actual financial statements, can and should be compared with the original pro forma statements to evaluate performance and measure success. In the first case above, continued operation of the four-store chain, the chance of significant variance from the pro forma statements is relatively small, if the assumptions are valid and pro forma statements were well prepared. All things considered, in many circumstances, it is relatively safe to assume that past performance patterns will continue. The likelihood of performance that is exceptionally good and the risk of performance that is exceptionally bad are relatively limited. In this case, the past is probably precedent to the future. So, if we had prepared pro forma financial statements estimating future performance of the four-store chain, when we compared actual performance expressed in *historical financial statements* with planned performance expressed in pro formas, it is more than likely that the historical financial statements pretty closely "tracked" the pro formas.

In the second case above, expansion of the chain to include a fifth standalone store specializing in higher-line merchandise, the likelihood of variance in actual performance from the pro forma statements is greater, regardless of how well prepared they were. If the underlying business premise is inherently good, the assumptions are valid, and the pro forma statements are well prepared, the opportunity for success may be great. These circumstances are what drive entrepreneurship. At the same time, if the business premise is fundamentally flawed or the pro formas are poorly prepared, the risk of failure may be significant.

Minimizing Risk with Careful Pro Forma Planning

Another way of looking at this is that good pro forma statements are ***"summary" planning tools***. They are one of the best ways that we can minimize risk of failing and maximize the probability of succeeding. In a sense, they crystallize all of the planning that we have done. They are our best expression of what we expect the future should be, or will be, for our business.

Therefore, the closer the pro forma statements are to the future reality, the better they will "predict" outcomes—profits or losses. Moreover, the better the quality of the prediction, the less likely we will face challenges that we are unprepared to handle. In a basic sense, it is easier to achieve success executing a sound plan than it is executing a weak plan.

Take a moment to consider "good preparation." In our simulation, our pro forma statements will be well prepared if we have taken every element of the business planning process into consideration in developing them. Really, every step of this book has required that we produce a "***deliverable***," a planning document such as a marketing plan, capital budget, or cash forecast. Individually, each of these documents must be based on facts, incorporate honest assessment of reality, and reflect best judgment.

The Importance of Consistency in Planning

Good preparation also means that each of our planning documents align or "tie" together correctly. For example, the elements of our capital budget should link directly with our cash flow statements and to our pro forma opening balance sheet. Similarly, sales estimates should tie to our cash flow statements and to our pro forma income statement.

In the case of a new business, like the one we have been simulating, virtually all of the conditions are hypothetical; the firm has no history or record of accomplishment against which to evaluate the validity of the pro forma statements. Therefore, there is a higher possibility that performance will deviate from plans and pro forma statements. In brief, there may be great opportunity for the business to exceed expectations. However, there may be equivalent risk that operations will fail to meet expectations.

Using Pro Forma Statements to Measure Future Success

Because the levels of risk involved in a new business are relatively high, pro forma statements take on a particular importance. They are possibly the most important set of documents—paper or electronic files—that you will prepare in the business planning and operational management processes. If you decide to proceed with the venture, in many respects, they will be the documented basis from which you launch it. Moreover, as you put your business plan into action, they will be the guideposts and benchmarks against which you will measure accomplishments, evaluate performance, and judge success.

They are important for many reasons. First, in assembling a set of pro forma financial statements, you are really integrating a lot of information that you have already gathered into a meaningful and easily understandable format. As you put the pro forma statements together, you are laying out an argument for yourself that the business is, indeed, a good venture, worthy of the risk to be undertaken. In reality, you are the most important user of pro forma financial statements. That is why you should be as thorough and objective as possible. As you prepare these statements, pause and take this opportunity to be your own best devil's advocate.

Well-Developed Statements Are an Important Tool to Secure Support

Second, in drafting pro forma statements, you are really reducing your comprehensive business plan, and all of its subsidiary parts, to a format that is meaningful to, and readily understandable by, the key people who are interested in your business. As we just noted, you are first in importance among interested parties. However, others, including lenders, suppliers, and potential investors, have critical interests in your new business plans. These interested parties will become your "sources of resources."

In all likelihood, these interested parties will evaluate your business plan in the context of competing alternatives. Investors, lenders, and suppliers will be looking at these statements, evaluating them not only for the reasonableness of the underlying business plan but also to assess the financial soundness of the investment proposal, the loan application, or the buyer-supplier relationship.

As you develop pro forma financial statements for your proposed business, it is important to keep sight of one fact. From a financial analysis perspective, neither you nor your potential investors, lenders, or suppliers have an historical track record to rely on. Moreover, while history is not necessarily a perfect predictor of the future, it does at least provide a basis of comparison and evaluation. Because your business has no history, investors, lenders, and suppliers will rely very heavily upon pro formas in making their decisions about doing business with you. This makes these statements extremely important.

Understanding Pro Forma Statements

A balance sheet is a *factual* representation of the financial condition of an existing business at an instant in time. A pro forma balance sheet is a *hypothetical* representation of the financial condition of a planned business at a future point in time. An income statement is a *factual* representation of the financial operations of an existing business during a specific period. A pro forma income statement is a *hypothetical* representation of the operations of a planned business during a future period.

Similarly, a cash flow statement may be prepared based on historical information or on a pro forma basis. Recall that cash flow forecasting was discussed in Step 10 and summarized in Table 10.2. The only difference between historical financial statements and pro forma statements is that the former are factual and the latter are hypothetical. The underlying mechanics of preparing them are much the same.

Income Taxes

In preparing pro forma financial statements for our simulation, we are going to make one major simplification: We are not going to take taxes into consideration. We will assume that tax preparation will be performed by our accountant, or another tax professional. While it is convenient for us to make this simplifying assumption, suffice it to say the effect of taxes in business planning are consequential and can influence the legal structure of the business that we select.

In a real-world context, it is quite likely that we would be preparing our business plan and pro forma statements in close consultation with our accountant or tax preparer. Throughout this simulation, we have encouraged the use of subject matter specialists and experts. Using the best professional talent and resources in business planning is critical to success. This is true for existing businesses and for new and emerging enterprises. Indeed, in the case of a new business, it could very well be the margin between success and failure.

Ownership of the Plan

One thing should be borne in mind. While we must work very closely with experts—accountants or tax preparers—it is critical that our plans and pro formas are "ours." In brief, we must fully engage in developing them, understand them thoroughly, honestly believe in their opportunity, and accept any risk that they imply.

Opening Pro Forma Balance Sheet

Let us turn to preparing an initial balance sheet for our simulation business. As we work through this part of the simulation, we will refer to several documents that we have already prepared. Remember, we are planning for a business that will generate gross sales of $330,000 during its first year of operation.

The information needed to prepare an opening pro forma balance sheet is found in Table 10.1, Capital Spending Plan, with respect to fixed assets, and Table 10.2, Cash Flow Forecast, with respect to cash revenues and expenses and loan payments. Please take a look at the Opening Pro Forma Balance Sheet in Table 11.1. Note that we estimate that we will need resources—total assets—of $100,000 to start the business. Some of these assets are characterized as "current" and others as "fixed." (See the CD-ROM for a blank version.)

Current Assets

We estimated that to start the business, we would need ***current assets***, assets that would be consumed in the course of the first year of operation, of $76,000. These assets include cash of $7,400. (Keep in mind that the term "cash" includes

TABLE 11.1 OPENING PRO FORMA BALANCE SHEET			
April 1, 2XXX			
CURRENT ASSETS		**CURRENT LIABILITIES**	
Cash	$7,400.00	Current Portion of Long-Term Debt	$5,730.00
Inventory	$48,000.00	Line of Credit	$-
Prepaid Rent-Security Deposit	$8,500.00	**TOTAL**	**$5,730.00**
Prepaid Electricity	$700.00		
Prepaid Insurance Deposit	$500.00	**LONG-TERM LIABILITIES**	
Prepaid Heating and Water	$300.00	Long-Term Debt	$64,270.00
Prepaid Marketing Expenses	$6,000.00	**TOTAL**	**$70,000.00**
Prepaid Travel Expenses	$2,500.00		
Security and Internet	$100.00		
Prepaid Accounting and Legal Services	$1,500.00		
Postage and Supplies	$500.00		
TOTAL	**$76,000.00**		
OTHER ASSETS		**OWNERS' EQUITY**	
Point-of-Sale System	$4,000.00	Owners' Investment	$30,000.00
Telephone and Fax Equipment	$500.00	Retained Earnings	$-
Leasehold Improvements Renovations	$5,000.00	Plus: Net Income (Profit)	$-
Furniture and Fixtures	$13,000.00	Less: Owners' Draw	$-
Store Sign	$1,500.00	**TOTAL**	**$30,000.00**
TOTAL	**$24,000.00**		
TOTAL ASSETS	**$100,000**	**TOTAL LIABILITIES AND OWNER EQUITY**	**$100,000**
CURRENT RATIO	**13.26**		
NET WORKING CAPITAL	**$70,270.00**		
DEBT TO WORTH RATIO	**2.33**		

exactly that—coin and currency, and deposits in checking accounts and other accounts that can be used essentially on demand.) Some of this cash would be required during the preopening period. Specifically, we expected that we would need $2,500 for buying trips to markets, $1,000 for legal services, and $500 to initiate accounting services.

As we noted above, we are estimating sales of $330,000 during the business's first year of operation. Because of the cost structure of the fashion industry, we believe that we will need an initial inventory of $48,000. We estimate this because, in this industry, cost of sales represents 51.3 percent of sales and rate of inventory "turn over" is approximately 3.5. That is, we estimate that the cost of goods sold will amount to 51.3 percent of $330,000, or $169,290. We also estimate that inventory will turn over, or be sold out completely, and have to be replenished, about 3.5 times during our first year of operation.

During the preopening period, we will prepay some items of expense. These include security deposits for space ($8,500), accounting and legal services ($1,500), advertising services ($6,000), electricity ($700), insurance ($500), utilities—heat and water ($300) and security and Internet services ($150). We will also buy postage and supplies for $500. All of these items will be classified as current assets in our pro forma opening balance sheet.

Other Assets

In starting the business, we will invest in some assets that will benefit operations over several years' time. In general, such assets are noncurrent and include "fixed assets" that will be consumed over a period of more than one year. Our investment in leasehold improvements—renovations of $5,000, furniture and fixtures of $13,000, point-of-sale system ($4,000), telephone and fax equipment ($500), and signage of $1,500—will benefit multiple periods and will not be used up within the year. For the sake of simplicity, we will assume that each of these assets will be used up through business operations over a five-year period.

Liabilities and Owners' Equity

Owners' equity is the amount of money that entrepreneurs have invested in the business—the amount that they have at risk. Liabilities are amounts of money that are lent to the business. It is important that the owners have a reasonable stake in the business to encourage lenders to put their funds at risk and for suppliers and vendors to take the risk of providing goods and services to an unproven entity. In this simulation, we are estimating that the owners are willing, and in a position, to invest $30,000 in the enterprise. With this investment, they will require additional funding of $70,000.

Just as assets are characterized as either "current" or "other" depending on the time frame over which they will be consumed by the business, financial "sources" are characterized as either ***current liabilities***, ***long-term liabilities***, or ***equity***. These classifications are based on the length of time that the sources will be available to the business. That is, current liabilities are expected to be paid off within one year, long-term liabilities to be retired over the term of the loan, and owners' equity available into the indefinite future—over the life of the business.

In our simulation, we are estimating that a bank will lend the new business $70,000 to be paid off over several years. We are going to assume that the bank will charge 10.5 percent interest on the loan. Therefore, for the first year of operation, the bank will charge $7,350 ($70,000 × .105) interest on the loan. For simplicity, we will assume that interest is based on the initial balance of the loan, and on the loan balance at the beginning of each year thereafter. The terms of the bank note call for payments of $1,090 per month, or $13,080 per year. This amount includes both principal and interest.

Therefore, in the first year of operation, of the $13,080 that we will pay the bank, $7,350 will be for interest on the loan and $5,730 will be applied to the principle of the loan—the balance that we owe. Keep in mind that the interest the business pays is an operating expense that will be deducted from revenue. The principle reduction is not. However, both interest expense and loan repayments are made from cash flow.

Non-Cash Expenses

There are some expenses that do not appear on the cash flow statement—***non-cash expenses***. These include ***depreciation*** and ***amortized costs*** that contribute to the operation of the business over a period of more than one year. In general, such assets are "expensed" over their useful lives.

As noted earlier in our simulation, "fixed assets" include leasehold improvements (renovations) of $5,000, furniture and fixtures of $13,000, signage of $1,500, point-of-sale system of $4,000, and telephone and fax equipment of $500. In the case of leasehold improvements, useful life is generally the term of the underlying lease, and their cost is expensed or amortized over that period. Furniture and fixtures are generally expensed or "depreciated" over an estimated period of time during which they are expected to be usable for the purpose for which they were required. Depreciable lives may be a function of declining physical utility, as when display racks simply wear out, or obsolescence, as when new generations of personal computer operating systems supersede earlier operating systems, bringing with them greater processing capability and increased functionality.

Note that there are many methods by which depreciation may be taken and costs amortized. These range from straight-line depreciation and amortization to accelerated methods, such as double declining balance depreciation and amortization. The former assume that assets will contribute utility to operations evenly over their useful lives. The latter assume that assets will contribute utility to operations more when they are new than as they age.

For the sake of simplicity, we are assuming that the useful lives of the leasehold improvements, furniture and fixtures, point-of-sale system, telephone and fax equipment, and signage are all five years and that their contributions to business operation will be evenly spread across this period. Therefore, we will include $4,800, one-fifth of the total cost of these assets, in projecting the first year's operating expenses.

But remember that while these are non-cash expenses, they are "real" expenses for the firm. For example, each year "fair wear and tear" will take its toll on furniture and fixtures such as display racks and counters, and technology, such as personal computers, will become obsolete. Because they are being consumed over time and through operation of the business over time, each type of asset will have to be replaced.

Pro Forma Balance Sheet Relationships

Lenders and suppliers are interested in the likelihood that the amounts owed to them, which are scheduled to be paid within 12 months, can actually be repaid. As we learned earlier, current liabilities are paid out of routine liquidation of current assets. The fact that we are estimating current liabilities of $5,730 and current assets of $76,000 means that we are "covering" current debt by a factor of 13.26 to 1. Net working capital, current assets minus current liabilities, is $70,260. This means that the business is very liquid, with a high probability of paying its current liabilities from orderly consumption of current assets through routine operation of the business. These conditions tend to bode well for successful operations. In the short run, they are very reassuring to suppliers and long-term lenders alike.

At the same time, one relationship—total liabilities, both long term and short term, to owners' equity—is approximately 2.33 to 1. This means that the business is modestly leveraged, that vendors and lenders have somewhat more at risk than the owners of the business do. The ***debt-to-worth ratio*** is good for a few simple reasons. First, the owners are willing and able to commit a significant share of the amount needed to start the business. Second, the principal asset of the business is inventory, and it turns over frequently, supporting a substantial level of sales despite a relatively small initial investment. Third, the business requires a relatively small investment in fixed assets—it is not capital intensive. If it operates successfully, the owners' investment in the business should increase, making the business less and less dependent on external sources for funding. This in turn can contribute to increased profitability, because declining debt will result in reduced interest expense. In general, lenders, especially those more aggressive and dynamic ones, are attuned to higher leverage in business start-ups and make such investments part of their overall business strategy. To mitigate some of the risk of this investment strategy, some lenders may seek the guarantee of federal, state, and municipal loan guarantees.

After we prepare a pro forma income statement for the first year of operation, we will "circle back" and prepare a pro forma balance sheet for the end of the first year.

Pro Forma Income Statement

As mentioned earlier in this step, an historical income statement is a factual representation of the operation of a business over a specific period. A pro forma income statement is the same, except that it is hypothetical. All of the information that you will need to prepare a pro forma income statement is found in the documents that you have already prepared.

Table 5.5, Monthly Sales Plan, provides for total sales of $330,000. Table 5.7, Break-Even Analysis, provided a gross profit percentage of 48.7, yielding a gross profit of approximately $160,710. From this amount, you must deduct operating expenses of $135,618. Deducting operating costs from gross profit yields net income of $25,092. See Table 11.2, Pro Forma Income Statement.

> *Note:* Table 5.9 does not take into consideration non-cash items such as depreciating and interest expense on loans. Also, Table 10.2 will differ in total operating expenses because depreciation and amortization are not taken into consideration on that table, while interest expense is there.

Take a moment to circle back to where we began, by comparing closing (Table 11.3) and opening balance sheets (see Table 11.1). See the CD-ROM for a blank version of Table 11.3. Table 11.4, Key Ratios, reflects a few interesting things. First, the debt-to-worth ratio increased slightly over the period. This is because while indebtedness decreased from $70,000 to $64,270, net worth declined at a greater rate from $30,000 to $20,092. Indebtedness decreased as a result of orderly repayment of debt from routine operations—a good thing! However, while the business "earned" $25,092, the owners withdrew $35,000.

TABLE 11.2 PRO FORMA INCOME STATEMENT	
April 1, 2XXX to March 31, 2XXX	
GROSS SALES	**$330,000**
COST OF GOODS SOLD	
Beginning Inventory	$48,000.00
+ Purchases	$164,000.00
- Ending Inventory	$42,710.00
	$169,290
GROSS PROFIT	**$160,710**
OPERATING EXPENSES	
Payroll	$26,200.00
Payroll Taxes	$3,668.00
Rent	$51,000.00
Utilities (fuel, electric, water)	$5,220.00
Maintenance and Repairs	$1,200.00
Security	$1,200.00
Telephone and Fax	$2,100.00
Interest on Loan	$7,350.00
Point-of-Sale System Maintenance/Updates	$1,200.00
Insurance	$2,080.00
Postage and Supplies	$2,400.00
Marketing and Advertising	$13,200.00
Travel and Entertainment	$5,000.00
Accounting and Bookkeeping	$4,200.00
Banking Services	$2,400.00
Depreciation and Amortization	$4,800.00
Miscellaneous	$2,400.00
TOTAL	**$135,618**
OPERATING PROFIT	**$25,092**

This means that the owners took out not only the entire business earnings but almost $10,000 of the amount initially invested in the business.

Note: Table 10.2 and 11.2 will vary in operating profit because 11.2 includes non-cash items such as depreciation while Table 10.2 does not. Also, Table 11.2 does not include principal payments on the loan, while Table 10.2 does.

TABLE 11.3 CLOSING PRO FORMA BALANCE SHEET			
March 31, 2XXX			
CURRENT ASSETS		**CURRENT LIABILITIES**	
Cash	$1,852.00	Current Portion of Long-Term Debt	$6,331.65
Inventory	$42,710.00	Line of Credit	$-
Prepaid Rent-Security Deposit	$8,500.00	**TOTAL**	**$6,331.65**
Prepaid Electricity	$700.00		
Prepaid Insurance Deposit	$500.00	**LONG-TERM LIABILITIES**	
Prepaid Heating and Water	$300.00	Long-Term Debt	$57,938.35
Prepaid Marketing Expenses	$6,000.00	**TOTAL**	**$64,270.00**
Prepaid Travel Expenses	$2,500.00		
Security Internet	$100.00		
Accounting and Legal Services	$1,500.00		
Postage and Supplies	$500.00		
TOTAL	**$65,162.00**		
OTHER ASSETS		**OWNERS' EQUITY**	
Point-of-Sale System	$4,000.00	Owners' Investment	$30,000.00
Telephone and Fax Equipment	$500.00	Retained Earnings	$-
Leasehold Improvements Renovations	$5,000.00	Plus: Net Income (Profit)	$25,092.00
Furniture and Fixtures	$13,000.00	Less: Owners' Draw	$35,000.00
Store Sign	$1,500.00	**TOTAL**	**$20,092.00**
TOTAL	**$19,200.00**		
TOTAL ASSETS	**$84,362.00**	**TOTAL LIABILITIES AND OWNER EQUITY**	**$84,362**
CURRENT RATIO	**10.29**		
NET WORKING CAPITAL	**$58,830.35**		
DEBT TO WORTH RATIO	**3.20**		

Second, while net working capital declined from $70,270 to $58,830, and current ratios declined a little, from 13.26 to 10.29, the business remains exceptionally liquid. Again, this is a good indicator to lenders and suppliers.

Third, the business proved exceptionally efficient from the owners' standpoint. That is, with the owners' equity of $25,046 on average [(beginning

TABLE 11.4 KEY RATIOS				
First Year of Operation				
	INCOME STATEMENT	AN OPENING BALANCE SHEET	A CLOSING BALANCE SHEET	
GROSS PROFIT RATIO				
Gross Profit	$160,710.00			
Sales	$330,000.00			
	0.49			
NET PROFIT RATIO				
Net Profit	$25,092.00			
Sales	$330,000.00			
	0.08			
CURRENT RATIO		**13.26**	**10.29**	
NET WORKING CAPITAL		**$70,270.00**	**$58,830.35**	
DEBT TO RATIO WORTH		**2.33**	**3.20**	
RETURN ON EQUITY				
Owners' Equity		$30,000.00	$20,092.00	
Average Owners' Equity				$25,046.00
Net Income				$25,092.00
				1.00
RETURN ON DEBT				
Debt		$70,000.00	$64,270.00	
Average Debt				$67,135.00
Net Income				$25,092.00
				0.37
RETURN ON ASSETS				
Assets		$100,000.00	$84,362.00	
Average Assets				$92,181.00
Net Income				$25,092.00
				0.27

owners' equity + ending owner's equity)/2)], the business earned $25,092, yielding a return on equity ratio of 1.00. This is a return on investment of 100 percent, an exceptional performance by any standard!

Another measure of efficiency is return on debt. That is, with an average debt of $67,135 [(beginning total debt + ending total debt)/2)], the business earned $25,092, yielding a return on debt of .37. This return on debt is very good, because the business is very profitable, but not heavily leveraged. Its success is not heavily dependent on the resources of others.

Finally, it is useful to consider return on assets. That is, with average assets of $92,181 [(beginning total assets + ending total assets)/2)], the business earned $25,092, yielding a return on assets of .27. This rate of return is very good also. It is so because a relatively small asset base is made up mostly of quickly turning inventory and very few fixed assets, and the profit margin on sales is high.

All of the business's ratios appear to be very solid in part due to its high rate of projected profitability. For the first year, we are projecting that the firm will have a gross profit margin of .49 and a net profit margin of .08.

If the owners can keep a bit more of profit in the firm in the future, it can increase the likelihood of success even more. Keep in mind that an "owner's draw" is the equivalent of the entrepreneur's salary. This is a critical point. In planning, we must consider that regardless of a new enterprise's profitability, the owner will continue to have personal living expenses. He or she must meet these personal expenses from some source—salary or draw from the new business, personal savings, income of a spouse or partner, or investment income, to name a few. If we are planning to start a business and do not have an alternative source of personal income, it is unrealistic to not project a salary from the new business. If the business will not support this, it calls into question whether or not we should proceed.

Points to Consider

Pro forma statements are estimates of the future. However, they should take into consideration as much factual information as possible. For example, once we have a particular Carytown property in mind for our store, it is far better to base pro forma statements on the financial terms of a specific lease than on average square-foot costs for leased retail space in Richmond.

We can be truly "fact-based" when we estimate lease costs for pro forma statements. It is harder to achieve this level of precision in estimating sales. We may have a lease, a contract, in hand for our store site. In addition, we may have contracts in hand for fashion jeans from our supplier. However, we have no contracts in hand for future sales of fashion jeans! It is often easier to estimate costs accurately than it is to estimate revenue. For this reason, we have to be particularly careful in developing sales estimates and to cross-verify and corroborate our estimates.

In planning, whenever we can, we should try to move from more general to more specific. The more specific we can be, the more likely our plans will pan out. Also, the more specific the plans, the greater the confidence we, our lender, and our suppliers can have in our plans. This confidence is critical at the outset.

As we operate the business, we will move from our first set of pro forma statements to statements of "actual" or "historical" performance. However, the process of planning does not stop, nor does the utility of pro forma statements diminish. As we conclude one operating period, we prepare estimates, projections, and "pro formas" of the next, and subsequent periods. However, now we will have some real history to help guide us. This history will help us understand where we were right, and where we were wrong. It will help us better prepare for the future.

Simulation

1. Prepare a pro forma balance sheet to reflect the store financial situation at opening (see Table 11.5 on page 282 and on the CD-ROM).
2. Prepare a second pro forma balance sheet to show projections for the end of the first fiscal year.
3. Prepare a pro forma income statement for the financial situation of the store at opening (see Table 11.6 on page 283 and on the CD-ROM).
4. Prepare a second pro forma income statement to show projections for the end of the first fiscal year of business.
5. Don't forget to write justifications for your statements.

TABLE 11.5 OPENING PRO FORMA BALANCE SHEET (blank)			
April 1, 2XXX			
CURRENT ASSETS		**CURRENT LIABILITIES**	
Cash		Current Portion of Long-Term Debt	
Inventory		Line of Credit	
Prepaid Rent-Security Deposit		**TOTAL**	
Prepaid Electricity			
Prepaid Insurance Deposit		**LONG-TERM LIABILITIES**	
Prepaid Heating and Water		Long-Term Debt	
Prepaid Marketing Expenses		**TOTAL**	
Prepaid Travel Expenses			
Security Internet			
Prepaid Accounting and Legal Services			
Postage and Supplies			
TOTAL			
OTHER ASSETS		**OWNERS' EQUITY**	
Point-of-Sale System		Owners' Investment	
Telephone and Fax Equipment		Retained Earnings	
Leasehold Improvements Renovations		Plus: Net Income (Profit)	
Furniture and Fixtures		Less: Owners' Draw	
Store Sign		**TOTAL**	
TOTAL			
TOTAL ASSETS		**TOTAL LIABILITIES AND OWNER EQUITY**	
CURRENT RATIO			
NET WORKING CAPITAL			
DEBT TO WORTH RATIO			

TABLE 11.6 PRO FORMA INCOME STATEMENT (blank)	
April 1, 2XXX to March 31, 2XXX	
GROSS SALES	
COST OF GOODS SOLD	
Beginning Inventory	
+ Purchases	
- Ending Inventory	
GROSS PROFIT	
OPERATING EXPENSES	
Payroll	
Payroll Taxes	
Rent	
Utilities (fuel, electric, water)	
Maintenance and Repairs	
Security	
Telephone and Fax	
Interest on Loan	
Point-of-Sale System Maintenance/Updates	
Insurance	
Postage and Supplies	
Marketing and Advertising	
Travel and Entertainment	
Accounting and Bookkeeping	
Banking Services	
Depreciation and Amortization	
Miscellaneous	
TOTAL	
OPERATING PROFIT	

STEP 12
Pulling It All Together

"Veni, Vidi, Vici" –*Julius Caesar, 47* B.C.

PROJECT OBJECTIVES

This step in the simulation will guide you through:

- Effective organization of your business plan
- Writing the summary/letter of intent
- Tabulation of the business plan
- Creating a business description and mission statement
- Distribution of the plan

Organizing the Business Plan

By this time the process of trying to open a business may have you running around screaming and pulling your hair out. If that is true, you might as well stop the process now and figure out what you really want to do with your life. We have stated on more than one occasion that opening and running a business is not for the faint of heart.

While the part about opening the business is just beginning, one step is close to being finished—the completion of the business plan. At this point all the really hard work has been done. You should have completed the following:

1. Evaluated yourself on your entrepreneurship, discovered where you can go for help, and developed an ***action plan*** (Introduction).
2. Determined the type of business to open, identified a ***target market***, and defined the business structure and its legal aspects (Step 1).
3. Decided where best to locate your business (Step 2).
4. Analyzed the various ways to fund your business, developed a ***sales plan***, and established realistic business goals (Step 3).
5. Prepared your business resume and a personal financial statement (Step 4).
6. Developed a ***profit and loss plan*** (Step 5).
7. Determined how to merchandise your store (Step 6).
8. Created a ***marketing plan*** for your business (Step 7).
9. Decided your personnel needs and how to manage them (Step 8).
10. Planned how to lay out your store, where to place fixtures, and how to display merchandise (Step 9).
11. Developed a ***capital spending plan*** and a ***cash flow*** forecast (Step 10).
12. Created a ***pro forma income statement*** and ***balance sheet*** (Step 11).

Congratulations! If you accomplished all of the above, then your business plan is almost complete. Organizing it into a cogent, easy-to-follow, professional-looking document is the next step.

Many references have been made to the U.S. Small Business Administration throughout this text as a premier source for business planning. The SBA's guide to organizing the business plan is well known, and using it is beneficial because almost all financial institutions recognize SBA's model. The model is found in the Introduction but is provided here again with some alterations to demonstrate the way that we recommend that you organize your plan. The SBA model can be found on the SBA's Web site at www.sba.gov.

Elements of a Business Plan

I. Cover sheet

II. Statement of purpose/summary/letter of intent

III. Table of contents

IV. Business description

V. Product/service

VI. Location

VII. Marketing plan

  1. Customer
  2. Competition
  3. Pricing and sales
  4. Advertising/promotions

VIII. Personnel/management plan

IX. Financial plan

  1. Loan applications (for entrepreneurs, not students)
  2. Sales plan
  3. Sales revenue monthly forecast
  4. Gross Margin Exercise
  5. List of fixed costs
  6. Pro forma profit and loss statements
     a. Three-year summary
     b. Detail by month, first year; by quarter, second and third years
  7. Break-even analysis
  8. Capital equipment and supply list (expenditures)
  9. Pro forma cash flow statements
     a. Three-year summary
     b. Detail by month, first year; by quarter, second and third years
  10. Income statement

11. Balance sheets
12. Justifications (assumptions) upon which all projections were based

X. A. Appendix items
1. Product/service: List of key vendors, detail of opening assortment
2. Location: Map of store area, floor plan with fixtures, completed lease agreement (or purchase agreement for entrepreneurs, if applicable), store fact list, floor plan with fixtures, store setup budget
3. Marketing: Marketing calendar
4. Personnel: Resumes and personal financial statements for all partners in the business, sample weekly work schedule, sample job descriptions for each different position in the business, organizational chart
5. The above items represent the minimum of entries in the appendix. You may choose to include other entries, such as photos of the store interior, pictures of representative products, or other documents that would help the reader to better understand your business. Just remember to evaluate carefully the entries in the appendix so as to not overstuff the section with unimportant information

B. Works cited
1. List of all books, Web sites, periodicals, and other reference material used in writing the plan
2. List of people interviewed in preparing the plan

C. Supporting documents
1. Tax returns of principals for last three years (for entrepreneurs only—not necessary for students)
2. For franchised businesses, a copy of ***franchise*** contract and all supporting documents provided by the franchisor
3. Copy of licenses and other legal documents
4. Copies of ***letters of intent*** from suppliers, etc. (for entrepreneurs, not students)

Using the above outline, organize the plan in a folder with easily identifiable tabs for each of the categories listed. We recommend either a three-ring binder or a soft folder with clamps to hold documents. Do not leave pages loose in a folder, as they may get out of order. Some people even elect to get the plan spiral-bound. The resulting document will not only look professional but it will also make a great first impression on those with whom you hope to do business. It may even be worth the extra expense to have the plan printed and produced by a professional publisher.

The Cover Sheet, Summary, and Letter of Intent

The cover sheet of the business plan folder needs only to include basic information. At a minimum, it needs to have the logo of the business, the name of the business, the name of the owner, and contact information. Many plans use the firm's business stationery for the cover sheet with prominent printing identifying the document as a business plan.

In writing the summary of your business or the letter of intent—actually these documents are synonymous—you should clearly state why the business is being opened, its highlights, its objectives, its mission, and the keys to its success. Bplans.com is an excellent source for information about business plan writing and formulation. The site also has many free examples of business plans and business plan summaries. Here is what www.bplans.com says about the business plan summary:

Writing an Executive Summary

One of the most important parts of a business plan is the ***executive summary***. The executive summary is also usually the first thing read and analyzed by potential investors, so it's a very important part of your plan.

What Should an Executive Summary Include?

Your executive summary should be no more than a page or two, and it should summarize all the other sections of your plan. It should include key financial numbers from your plan, as well as brief summaries of other important sections.

When Should I Write an Executive Summary?

It should be the last thing you write in your business plan, even though it's usually the first thing read by others. Its concise length and summary format will enable the reader to quickly understand what you plan to do with your business.

Banks and ***venture capitalists*** are busy people and if their interest is not piqued at the start, they will not continue to read your plan. VCs receive thousands of plans, many of them the size of a book. They will only read a plan if the first few pages indicate that it is worthy of further exploration, which is why your executive summary is so important.

The Business Description

The purpose of the Business Description section is to entice the reader and to prompt him or her to review the entire plan. This is the entrepreneur's opportunity to state the purpose and highlights of the plan. This section should address the following issues in a succinct summary:

- Reasons for starting the business
- Summary of the industry: status, predictions for continued health, key trends, how your business will fit into the industry, what opportunities exist locally for growth in your industry, and what challenges the industry faces now and in the near future
- Target customers
- Products and services
- Store ***niche***, why the store will be successful and profitable
- Legal business form: sole proprietorship, partnership—be specific
- Business type: retailing, service, manufacturing

- Business status: new, expanding, buying an existing business, franchise
- Is the business currently funded? What kind of financial assistance do you seek? How will the money be used?
- Goals and objectives of the business
- Mission statement

The reason for writing the business description as one of the last parts of your business plan is that it is a summary. By the time that you write it, you will have formulated your plans, evaluated your ability to be profitable, and thoroughly researched your market, product lines, and target customers. Potential problems would have been identified and solutions reached. With the completion of financial documents, the loan amount needed to open the business would have been calculated. In other words, at this point, you know a great deal about your business and you have obviously determined that it is a viable business. With this knowledge and confidence, you can successfully write a Business Description section that will grab the reader's attention and keep him or her interested.

This section should not be too long—keep it short enough to intrigue the audience to read on—but long enough to address the issues. Also, keep it positive; this is not the time to detail all the things that can go wrong with the business. You dealt with possible problems when you wrote the Personnel section; you don't need to repeat them here. To help you organize your thoughts for this section, we suggest that you start with a list of the most important and most positive aspects of your plan, rate them in order of priority, and then proceed to write.

The Mission Statement

It is important that every new business have a mission statement to describe the purpose and goals of the business. Writing a mission statement will encourage the new owner to identify in a concise manner what he or she wants the business to be: What will be the values, purpose, image, and policies?

In addition to helping the owner explain his or her purpose, the mission statement can be used to inspire and educate sales associates about the company and to give them a framework to use in serving customers and promoting goodwill for the business. Some owners elect to put the store mission statement onto a poster and hang it in a storeroom or break room so that employees will see it often and remain aware of its importance to the success of the business.

Your mission statement should provide the basis upon which the marketing campaign is built. It should be uppermost in the mind of the entrepreneur when he or she is planning any marketing, advertising, new policy, visual merchandising program, or even store expansion. In Step 7 we discussed the brand promise, including the mission statement that we built (with the able assistance of our business consultant, Michael Sisti of Sisti and Others) for our simulated store. It states quite succinctly our goals, brand image and promise, and intentions for our new store. Referring back to it will help us to stay focused on our goals. For the business description section of the business plan, write your mission statement that projects your image and clarifies your purpose.

Distribution of the Plan

Now that the plan is completed, the entrepreneur can decide to whom it should be distributed. He or she may elect to show it to the following people:

- Bankers
- SBA officers
- SCORE volunteers who gave assistance along the way
- Lawyer
- Accountant
- Venture capitalists or ***angel investors***

- Important potential clients
- Key ***vendors***
- Key employees
- Mom and Dad

Just kidding about that last one, but surely everyone who does all this work will want to show it to some family members or friends who might give last-minute encouragement and advice.

When submitting the plan to anyone whom you desire to help finance your business, a separate targeted letter should be written to include with the plan. This letter should be personalized and state specifically what you are asking the reader to do. If you hope that the reader will invest money in your business or grant a loan, then state that in the personalized letter.

The smart entrepreneur will chose wisely among the list of possible people to whom he or she shows the plan. He or she will not want an unfriendly reader to either attempt to thwart plans or copy ideas before they are fully implemented.

Simulation

1. Complete the executive summary for your plan.
2. Write the business description for your plan.
3. Pull all parts of the plan together as outlined above, being careful that you have all necessary financial forms (complete years two and three of the P & L and cash flow forecast if you have not previously done so).
4. Put your plan into a professional format with tabs to separate the sections and place into a binder or other enclosure as described above.
5. Be sure that you have signed your lease agreement and that it has been signed by the landlord (students may simulate this part). Also, sign your executive summary in the front of the plan.

CONCLUSION
Managing Your Success

"Victory belongs to the most persevering." —*Napoleon Bonaparte*

PROJECT OBJECTIVES

This step in the simulation will guide you through:

- Effective customer service
- Maintaining a profitable business
- Using technology
- Evaluating success and failure
- Benefit of professional and business organizations
- Planning for emergencies

Here we are—at the end of the beginning. By now you will have worked your way through the myriad details of planning and opening a business. It is a tedious process and not one for the faint of heart (Figure C.1). Heed the quote from Napoleon at the beginning of this step. Remember, perseverance is the businessperson's best friend.

Opening a business is one thing. Staying in business is an entirely different prospect. While it is not good to dwell on the negative, you must remember that most businesses do not succeed or they close prematurely. According to the U.S. SBA, the following are the most likely reasons for business failure:

- Lack of experience
- Insufficient capital (money)
- Poor location
- Poor ***inventory management***
- Overinvestment in fixed assets
- Poor credit arrangement management
- Personal use of business funds
- Unexpected growth
- Competition
- Low sales
- Poor record keeping

Figure C.1 *When you own your own business, there are often many nights spent working late.*

Following the guidelines in this text will help minimize the above pitfalls and give your business a better than even chance to succeed. Being in business is a continuous, labor-intensive, midnight-oil-burning, hair-pulling-out, stress-inducing process. The often-quoted comment that people go into business to quit working 8 hours a day for somebody else so they can then work 24 hours a day for themselves is more than appropriate. Owning a business is indeed a 24-hour-a-day job. As long as you are going into the venture with both eyes wide open, have done your homework, know your strengths, and identified your weaknesses, your chance of success is greatly improved.

The Future

Now that you are in business, it is presumed that you will want to stay in business. Don't laugh. Many people go through all of the tedious processes involved in opening a business and find that it is just not for them or is not what they dreamed it would be. If you are lucky, you will find out early on in the process whether or not being in business is for you. For some people it takes a lot longer and a lot of money to make that determination.

Are you one of those people who cannot wait to get up in the morning so that you can open your businesses doors? If the answer is yes, then you need to now concentrate on the ways to help you keep those doors open. Nothing can be more satisfying after achieving your goal of opening a business than seeing the business thrive.

The surest most successful thing that a business owner can do to keep the business humming along is to maintain effective customer service. Customers are the mother's milk of all businesses, and one unhappy customer can be devastating to a business's reputation. It cannot be reiterated enough times: *Make the customer happy!*

Ask yourself the following questions: How many times do you think about seeking out the store manager when a salesperson gives you great service? Now, how many times do you think about complaining to a store manager when you get lousy service from a salesperson? Get the point? People are more likely to complain about bad service than they are to compliment good service. The following sections explain tips regarding good customer relations.

Know Your Customer

Rhonda Adams, president of The Planning Shop and author of *The Successful Business Plan: Secrets and Strategies*, writes the following in a *USA Today* article (http://www.usatoday.com/money/smallbusiness/columnist/abrams/2005-02-18-persona_x.htm):

> As small business owners, we pride ourselves on providing great products and go-the-extra-mile customer service. We know those are the keys to our survival. But how can we ensure we're really satisfying our customers? How can we make our products even better?
>
> Although we think we know our customers well, the reality is we probably don't. Most of us, after all, develop our products or services because we ourselves recognize a need in the market. I know dozens of entrepreneurs who started their businesses because there was something they wanted to buy that just didn't exist.
>
> As a result, many of our products or services tend to reflect our own interests, needs, and abilities. Those aren't necessarily the interests or needs of our target customers. In our rush to get a product out the door or to get our company up and running, we don't have the time—or money—to do a lot of market research.

Ms. Adams has hit the nail on the head. Businesses are started by people who see perceived needs. Those needs are most often the business owner's.

When you get into business, you may find that the perceived need fills the niche you had in mind, but in order to keep customers, it is necessary to expand that need or to modify it.

Become a Destination Shop

Becoming a ***destination shop*** is one of the more difficult things a business can accomplish. Not only do you have to have a unique product but you also have to develop a sense among your customers that you are the only place to go to get what it is that they want. Developing that almost intangible sense among customers often takes a lot of years and many tests and trials of the right mix of products and salesmanship. However, once established, that goodwill among your customer base is money in the bank (Figure C.2).

Many things go into making a business a destination shop. According to an article in *Main Street, Beaufort, USA*, "The owner must operate, market, and position the business in the customer's mind. In other words, the consumer would be insane not to choose that business for their purchases." Great examples of businesses that have successfully branded their products and are successful destination businesses are FedEx, Nike, Disney, Victoria's Secret, Microsoft, McDonald's—the list goes on and on. You will also understand by looking at these companies that it is the product that makes the business a destination, not the physical location. The product that they sell and the reputations that they have developed in standing behind those products are what makes them the household names they have become.

Use Value-Added Strategies and Go the Extra Mile

Value-added strategies come in all shapes and sizes, but for the small retail store, anything that makes the customer want to buy in your store is value added. Value added in the business sense means to add something to your product or service that makes it more valuable to the customer. Making

Figure C.2 *Customer loyalty is key to a successful business.*

products more valuable to the customer means more profit for the business.

According to Adam M. Brandenburger and Barry J. Nalebuff, coauthors of *Co-opetition*, "To create value, a business needs to align itself with customers, suppliers, employees, and many others. That's the way to develop new markets and expand existing ones. But along with creating a pie, there's the issue of dividing it up. This is ***competition***. Just as businesses compete with one another for market share, customers and suppliers are also looking out for their slice of the pie. Creating value that you can capture is the essence of business."

Some simple examples of value-added strategies follow:

- Partner with other businesses to put discount coupons in each other's store.
- Partner with other businesses to buy advertising.
- Offer discounts for customers that buy online.
- Have free coffee in your store for your customers.
- Offer free delivery for your customers.
- Have a generous return policy.
- Offer free gift-wrapping.
- Offer a bridal or groom registry.
- Send out regular e-mails to customers advising them of fashion trends or the latest fashions.
- Send cards to customers on their birthdays or anniversaries along with a discount coupon.

Most of the above suggestions will cost little or no money and will result in enormous benefits in the form of loyal and satisfied customers. Remember, you want your customers to be telling their friends about your business and how great it is. Anything you can do to encourage good customer relations is going to pay off in increased profits.

Use Relationship Marketing

Wikipedia defines relationship marketing as a "philosophy of doing business, a strategic orientation that focuses on keeping and improving relationships with current customers rather than on acquiring new customers. It is the use of the wide range of marketing, sales, communication, and customer care techniques and processes to identify your named individual customers and create a relationship between your company and these customers."

At the heart of relationship marketing is the development of a strong bond—even friendship—between businesses and their customers. That bond creates such a strong relationship that the customer would not think of going elsewhere to buy from another business. Businesses, of course, would kill to have those kinds of customers because studies have shown that retaining an old customer is only 10 percent of the cost of obtaining a new one.

According to Buchanan and Giles, increased profitability associated with customer retention efforts occurs because of the following:

- The cost of acquisition occurs only at the beginning of a relationship, so the longer the relations, the lower the amortized cost.
- Account maintenance costs decline as a percentage of total costs (or as a percentage of revenue).
- Long-term customers tend to be less inclined to switch and also tend to be less price sensitive. This can result in stable unit sales volume and increases in dollar-sales volume.
- Long-term customers may initiate free word-of-mouth promotions and referrals.
- Long-term customers are more likely to purchase ancillary products and high-margin supplemental products.
- Customers that stay with you tend to be satisfied with the relationship and are less likely to switch to competitors, making it difficult for competitors to enter the market or gain market share.

- Regular customers tend to be less expensive to service because they are familiar with the process, require less "education," and are consistent in their order placement.
- Increased customer retention and loyalty makes the employees' jobs easier and more satisfying. In turn, happy employees feed back into better customer satisfaction.

Maintaining a Profitable Business

The reason that you go into business is to satisfy a dream, and you want that dream to thrive. Unless you are independently wealthy or have a rich uncle who indulges you, your business is going to have to make a profit in order for it to survive. To make a profit, a business must control costs. Controlling costs means maintaining accurate records.

By now you should have in your possession the one document that gives you a head start in maintaining accurate records and controlling costs. That document is your business plan. The business plan contains all of the data you need to keep track of your business and ensure that it is meeting its stated goals.

The SBA tells us that as a management tool, the business plan helps you track, monitor, and evaluate your progress. The business plan is a living document that you will modify as you gain knowledge and experience. By using your business plan to establish timelines and milestones, you can gauge your progress and compare your projections to actual accomplishments.

As a planning tool, the business plan guides you through the various phases of your business. A thoughtful plan will help identify roadblocks and obstacles so that you can avoid them and establish alternatives. Many business owners share their business plans with their employees to foster a broader understanding of where the business is going. It should be constantly evaluated and updated at least once a quarter.

Another factor that will help maintain business profitability is for the

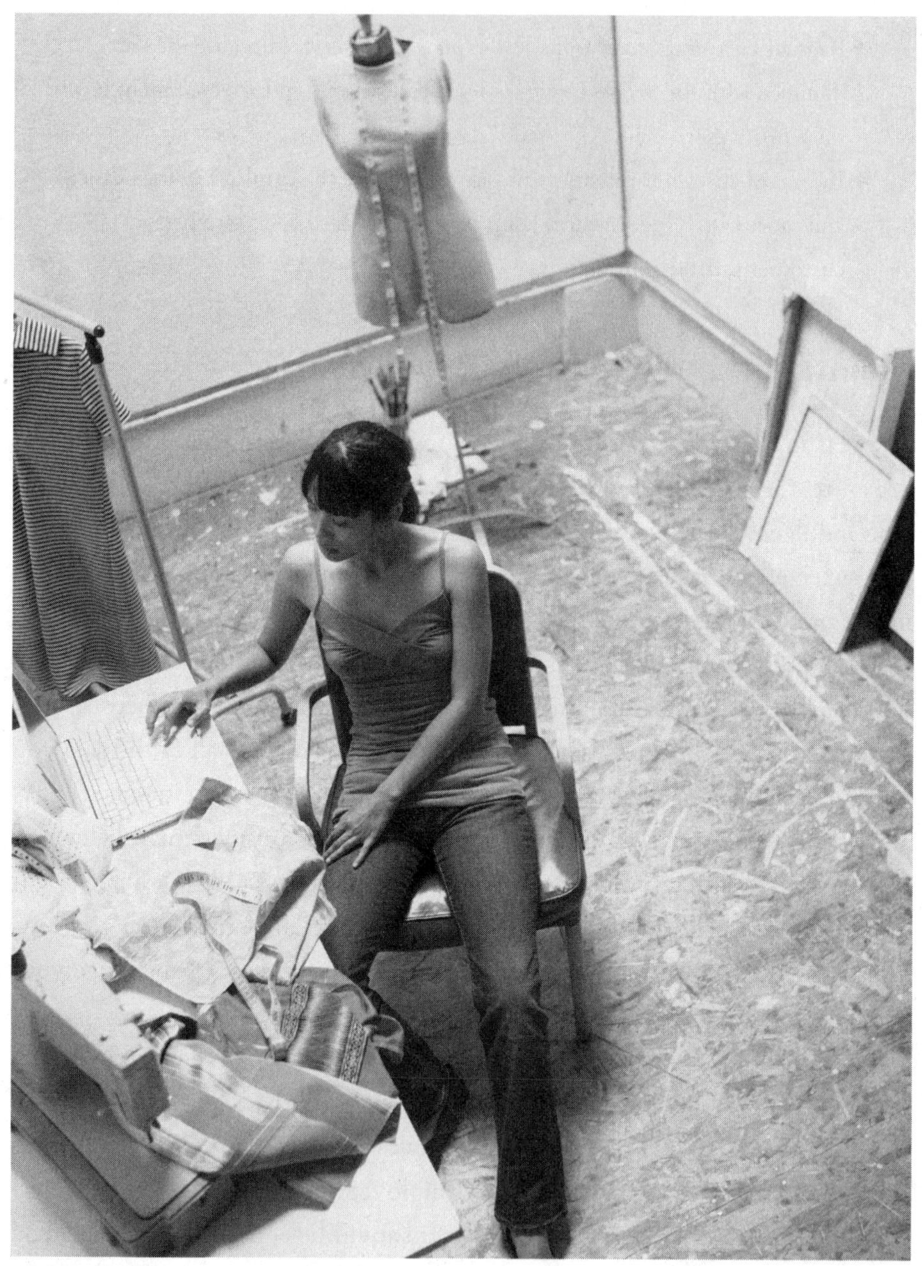

Figure C.3 *Keeping up with industry trends via the Internet helps a business maintain profitability.*

business to keep up with industry trends. In the fashion world, trends are constantly changing and evolving. The bread and butter of the fashion industry are the changing styles that help generate demand. Along with the renewed demand of the latest styles in clothing fashion are the accompanying accessories—shoes, jewelry, handbags, and so on.

Not only does the business need to keep track of product trends but also the trends in technology. The Internet in recent years has exploded on the retail scene, and sales via online purchases have skyrocketed. Businesses that ignore the power of the Internet and how it relates to them do so at their own peril. Note that the Internet is the best source for quickly evaluating the trends that relate to your business (Figure C.3).

Once your business is established and humming along, inevitable questions will come to mind: "Should I expand?" "Should I not expand?" "If I decide that expansion is the correct decision, how do I go about it?" The decision to take your business to the next level is obviously important and not to be taken lightly.

Phil Holland, founder of MyOwnBusiness.org, lists the following top 10 don'ts in considering whether or not to expand your business:

1. Don't think about your second store before the first one is reliably profitable.
2. Don't sign long-term leases (instead ask for short-term plus options).
3. Don't open the second store before delegation controls are in place.
4. Don't let self-confidence overcome prudent and calculated decision making.
5. Don't give your personal guarantee wherever asked.
6. Don't let commissioned salespersons set prices.
7. Don't fail to promptly cut costs to maintain positive cash flow.
8. Don't pursue a "commodity" business (one without pricing power).
9. Don't cut the value or quality of your product or service.

10. Don't delegate signing checks (any amount) or making capital expenditures.

It is also necessary to understand that business expansion does not necessarily mean renting another store. It can mean expanding the square feet of your existing location. It can mean expanding your product line. It can mean expanding your store hours. It can mean creating a larger online presence. It can mean hiring more employees. Just be sure that any decision to change your business model is taken with great care.

Technology and Your Business

At the center of every successful business are the software systems used to process information. Isis, Inc. is a business technology solution provider located in Richmond, Virginia, and its account executives help small businesses with their technology problems. The firm lists 90 different technology applications that businesses can use to enhance their use of technology. The following is a sample of the technology help that Isis provides:

- Accounting and finance systems
- Order processing systems
- Customer relations management
- Time and billing systems
- Service systems
- Human resources
- Network integration
- Directory design and implementation
- E-mail
- Internet access

- Web site hosting
- Web site maintenance
- Facilities management
- E-commerce development
- System performance optimization
- Financial report writing
- Training
- Business continuity and disaster recovery
- Strategic planning

According to Isis, staying ahead of your competition requires the ability to foresee, embrace, and innovate in the face of change. A business's technology vision must work with its business vision to help achieve business goals. Many companies do not properly integrate these two, making it more difficult for their businesses to succeed.

For a very small business, a grandiose technology system may be too expensive, but companies like Isis can design systems both large and small. An initial technology investment will include point-of-sale equipment that will no doubt include a cash register and perhaps credit card machinery along with the necessary software. In addition, the firms need security and fire protection. Phones, pagers, fax machines, and printers are other items that will be required purchases. At least one computer will be needed and programmed to integrate all of the technology the business will be using. The computer will also be used to track sales and to order merchandise.

Bill Gates of Microsoft has reportedly said that there will eventually be only two kinds of businesses: those that use the Internet for business and those that aren't in business at all. Change is tough, but those businesses that can manage it instead of being frightened by it will be in front of all of those businesses that won't or can't adapt to the changes that technology presents. Today, businesspeople are often asked for the address of their firm's website. If you don't have one, then that potential customer will likely dismiss you as someone with whom he or she will do business.

Having a Web site for your business is a very easy and a fairly inexpensive proposition. There are plenty of companies, such as Isis mentioned above, that provide Web site design services. These firms will design and maintain your site for a fee. If you want to save money, you can go to your local business development organization and ask for a referral to a Web site designer. Many times students who are very Internet-savvy can help design a Web site for minimal cost. Professional and trade organizations often will assist members with technology needs or provide them at minimal cost.

There are many benefits to having your own Web site:

- Reaching thousands, if not millions, of customers quickly and inexpensively
- Selling products and services easily
- Online publishing
- Leveraging advertising dollars
- Updating information quickly
- Bringing more people into your stores
- Getting you a foothold in cyberspace
- Making business information available
- Quickly answering frequently asked questions (FAQ)
- Allowing quick customer feedback
- Allowing 24-hour service
- Helping streamline information for employees, especially if you have more than one site or have employees on the road
- Opening up new markets and reaching specialized ones

According to Net Access, an Internet service provider, "The population on the Internet is well-educated and affluent. Most own a computer; others have access to one. Internet users, like most people, are interested in convenience. Many prefer the ease of finding services or shopping and ordering directly from their computer screens."

For our simulated store, we created an information-only Web site. We referred to this Web site as a part of your marketing plan, which was developed in Step 7. We also included the cost of developing and maintaining this Web site in our marketing calendar.

Evaluating Business Successes and Failures

Throughout this text, we have described why businesses succeed and why they fail. We have repeatedly stated that going into business is not an easy proposition, and if you have not paid attention when it was stated earlier, then please pay heed now (Figure C.4). The heartache of a ***foreclosure*** on your mother's home that she used to back your venture or having to deal with ***bankruptcy*** filings is too much to bear. Please, please, please make sure that if

Figure C.4 *A good understanding of why businesses fail and succeed is important when starting your own business.*

you want to be in business you have done everything in your power to make it a success.

On the other hand, there is nothing more uplifting than seeing a dream take flight and flourish. To know that an idea you had is now making others happy is indeed a euphoric feeling. The fact that you are contributing to your community by employing people and providing a service or product that people want is very uplifting. However, the very best feeling is a pride that comes from accomplishing something that has always been your dream.

According to Jan King, author of *Business Plans to Game Plans: A Practical System for Turning Strategies into Action* (John Wiley & Sons, 2004), the top 10 reasons that businesses succeed are that they have the following:

1. The experience and skills of the owner and managers
2. The energy, persistence, and resourcefulness (the will to make the business succeed) of the owners. Don't give up
3. A product that is at least a cut above the competition and service that doesn't get in the way of people buying
4. The ability to create a "buzz" around the product with aggressive and strategic marketing
5. Deal-making skills to sell the product at the highest possible price given your market
6. The ability to keep developing new products to retain and build a customer base
7. Deal-making skills to work with resource suppliers to keep costs low
8. The maturity to treat employees, suppliers, and partners fairly and respectfully
9. Superior location and/or promotion, creating a connection between your product and where it can be obtained
10. A steady source of business during both good economic times and downturns

Businesses fail for many reasons. As a business owner, you must also be aware that circumstances sometimes dictate that a business must be shut down or closed. Recognizing this fact is necessary so that the closure will be less painful to the owner, the employees, and others who count on the business. The SBA has excellent information available at its Web site (www.sba.gov) that will assist the owner in this difficult process.

Belonging to Professional and Business Organizations

Americans are joiners. Just look at the number of clubs, societies, orders, fraternities, sororities, and unions that exist in this country. Obviously there are many such groups dedicated solely to business.

Belonging to such organizations on behalf of your business can be very beneficial. Groups like merchant associations and chambers of commerce are terrific not only for meeting your business contemporaries but as great sounding boards for help and advice on how to conduct your business. They also provide much of the statistical material you might need in deciding where to locate your business or what market to pursue.

Benefits of belonging to business organizations include the following:

- They allow you to interact with like-minded people.
- Many offer members discounts on products or services they provide.
- Many offer free advertising in the organization's publications.
- Many provide training opportunities for members.
- Belonging to such groups increases your business's credibility.
- Industry-specific organizations help you keep track of advances in your field.
- Belonging to nonspecific organizations can be eye-opening and can give you insight that you would not get otherwise.
- They provide priceless opportunities for networking.

These same reasons apply to subscriptions to business publications. If you are in the fashion industry, then a subscription to *Women's Wear Daily* is a must. In the same light, if you are in the financial industry, then you no doubt will read the *Wall Street Journal* daily, although many businesses subscribe to the *WSJ* because of its multifaceted coverage of just about all business topics.

Planning for Emergencies

No one can foresee what is around every curve in the road. The prudent businessperson will do his or her best to plan for unforeseen events. A dip in the stock market, a terrorist attack, a natural disaster like a flood or hurricane, a store fire, or the sudden death of a partner or critical employee can have a devastating effect on a business (Figure C.5).

Figure C.5 *Planning for emergencies and unforeseen disasters is necessary for all business operations.*

Insurance can help alleviate some fears, such as destruction to business property from a fire or a storm. Insurance costs vary widely, and the business owner needs to shop around to get the best deal available. Note that flood and hurricane insurance are very expensive, and many financial institutions will not loan money to businesses in areas prone to those kinds of natural disasters.

It is also possible that if the store premises are leased, the building owner will cover that damage to the structure or the costs will be prorated in the leasing arrangements. Having a reliable and trustworthy insurance agent to give the business owner advice is the best solution for any insurance questions. Professionals should also address questions of security. The landlord may also have an alarm system as part of the lease agreement.

If at all possible, the business should have built-in financial reserves to help it weather storms that could be both literal and actual. Small business start-ups experience the most problems in having adequate financial reserves on hand because they generally have maximized their financial resources in getting the business started.

If a financial crisis was to occur and the business or owner has little or no funds to cover them until the crisis is over, here are some steps that can be taken:

- Go immediately to any creditors and tell them of your situation. They do not want to put you in default, and the great majority will help you work out a new payment schedule or give you new terms.
- Reduce all costs. Adjust the hours that the store is open, reduce employee hours, lay off employees, and so on.
- Get the business open and running at full speed as soon as possible.

The best way to deal with emergencies is to have a plan in place. Do your best to identify the kinds of things that can happen and address how you will respond. When an incident takes place, you will then be much better prepared to deal with it. Take a look at Entrepreneur.com. Their Web site gives great advice on how to deal with actual emergencies (http://www.entrepreneur.com/management/operations/article78212.html).

Simulation

1. Wrap up your hopes, dreams, and business plan and find financial backing.
2. Buy the product.
3. Hire the people.
4. Open the store.
5. Make millions and . . .

"Live long and prosper." –*Mr. Spock,* Star Trek

Appendix

Business Plan

BELLISSIMA

5710 Patterson Avenue · Richmond, Virginia 23226
Tel: 804.555.6730 · E-mail: info@5710bellissima.com

Owners:
Kate Aliberti
Samantha Baum

Business Plan Contents

EXECUTIVE SUMMARY

BELLISSIMA

5710 Patterson Avenue · Richmond, Virginia 23226
Tel: 804.555.6730 · E-mail: info@5710bellissima.com

Ms. Rachel Schnaier
SunTrust Bank
919 East Main Street
Richmond, Virginia 23219

Re: Loan Request for $40,000

Dear Ms. Schnaier:

With five years of working in the fashion industry, we are requesting a SBA Guaranteed Community Express loan in order to open our own dress boutique, named Bellissima.

Current trends in society necessitate a boutique that caters to the social, professional, and everyday occasion fashion needs of Richmond women. No store in the Libbie and Grove Shopping Area has the wide variety of dress types with the additional stylist services that we will offer. These women have the financial stability and disposable income to make our store profitable. Our mission and goal is to provide customers with a wide variety of fashionable dresses in a chic but comfortable atmosphere where they will receive exceptional customer service and have additional stylist services. This will be accomplished by our dedication and love for fashion combined with our high standards of excellence.

Our education and experience of the fashion industry include working at many different retailers with a wide variety of merchandise and sizes along with graduating with degrees in fashion merchandising. This gives us a strong, well-rounded knowledge of the fashion industry in every aspect. Our research and financial forecasts show that investing our $22,000, combined with a single $40,000 loan, will give Bellissima the funding to successfully achieve our goals, with the ability to pay back the loan in full in only three years.

The funds we wish to borrow will enable us to open our business, initially allowing us to purchase our opening inventory, fixtures, and promote Bellissima. We plan on opening our store September 1, 2008. Granting us this loan will allow us to fulfill our goal to make the Richmond woman feel confident and special, and look stunning.

Our business plan is attached, in which you will find all the information you will need. If you have any questions or need more information, please contact Samantha Baum at (804) 555-0582.

Thank you for your consideration.

Samantha Baum Kate Aliberti

Located on the corner of Libbie and Patterson Avenues, Bellissima offers a variety of exclusive dresses, handbags, and jewelry for the trendy woman in a comfortable setting. Exceptional customer service and a personal stylist will make finding the perfect dress for any occasion fun and easy, whether in-house or around town.

Bellissima's target customer is a woman between the ages of 18 and 45 who already shops in the Libbie and Grove Avenues area. Our target age range overlaps with the existing shopper of the area, making our store a perfect fit. Bellissima's target customer is more fashion-conscious than the average woman, selecting trendy garments in addition to wardrobe staples. She desires exclusive, well-made pieces that will offer her a unique look compared to other women. Exclusivity of merchandise will satisfy the trendy customer's need to feel distinctive and fashionable.

A unique-to-the-area service that we will offer is our in-house stylist. The stylist can work with customers both in and outside of the store. Customers in the store will receive complimentary stylist assistance, including a fashion consultation and Polaroid photographs of possible outfits suitable for any upcoming social occasion. For a fee, a Bellissima stylist will accompany a client on a shopping trip, or to "shop" her closet at home for new looks.

Being a boutique store, we will be able to distinguish ourselves from bigger stores through exceptional customer service—going "above and beyond" what large, mass retailers can offer. Our pleasing atmosphere will entice customers by stimulating all five senses and creating a total shopping experience.

Bellissima will be a partnership between Samantha Baum and Kate Aliberti. Both women have a combined expertise that will enable them to successfully open and manage a small retail boutique. Being familiar with utilizing weekly sales reports to track, analyze, and communicate business results and determine strategies will enable the pair to maximize sales.

Bellissima's mission statement is simple: ***Dress to impress.***

Offering a variety of types of dresses matched with complementary handbags and jewelry shows the store's commitment to making every woman look both beautiful and fashionable.

PRODUCT AND SERVICE

Business Description

Bellissima's business will be twofold: selling a variety of dresses and providing the additional service of an in-house stylist. Bellissima will offer a high level of customer service to complement our superior selection and styling assistance.

Store Merchandise

Bellissima will offer a variety of dresses, handbags, and jewelry for the trendy woman aged 18 to 45. Our store will give customers a variety of types of dresses: cocktails, special occasions, and party dresses; day/sun dresses; and work-appropriate dresses in specialty and regular sizes. We will offer handbags that complement the stock of dresses, ranging from evening bags to clutches to everyday handbags. We will offer costume and semiprecious earrings, bracelets, rings, and necklaces that will complement dresses in stock.

In-House Stylist

A unique-to-the-area service that we will offer is our in-house stylist. The stylist can work with customers both in and outside of the store. Currently, only one other personal stylist, a freelancer, exists in Richmond, so our stylist facet will add to our store's niche. Customers in the store will receive complimentary stylist assistance. This will include a fashion consultation, plus Polaroid photographs of possible outfits suitable for their occasion. For a fee, we will extend this service to assisting the customer in her chosen location, either in other stores or at her home. Building on the at-home aspect, our store will offer a closet consultation to help customers organize their closets, streamline their closet through elimination of old styles and suggestion of needed pieces such as accessories for each outfit, and rethinking existing clothing to create new outfits. Along with this, we will create a "look book" of her wardrobe and its possible outfit combinations. To help bring attention to this service, we will have a sample look book that is kept in a high-traffic area in the store.

Customer Service

Being a boutique store, Bellissima will be able to distinguish ourselves from bigger stores through exceptional customer service—going above and beyond what large, mass retailers can offer. Upon entering Bellissima, customers will be greeted and offered a beverage of either fresh-flavored mint or lemon water, sparkling water, or white wine if age-appropriate. Children will have a selection of toys and coloring books to occupy them. A cozy seating area, near the dressing rooms, will provide a variety of magazines and a comfortable place to sit for those accompanying the shopper. A sales associate will then individually work with the customer to put together an appropriate outfit while offering fashion advice and showing the customer new merchandise.

Each dress will be offered in four sizes, in order to create exclusivity and assure the shopper that her outfit will be unique at any social function. Our sizes will range from 0 to 8, with the offer of special order for sizes not offered in the store. For those that spend over $500 on their visit, the associate who assisted them will send a handwritten thank-you note to let the customer know how much we appreciate their business. By writing notes, we aim to build on customer loyalty and create a personal relationship between the customer and the store.

Atmosphere

Bellissima's atmosphere will entice customers by stimulating all five senses and creating a total shopping experience. Upbeat music will play in the background to put the customer in a positive mind-set. Taste will be satisfied with a customer's beverage of choice. Light and fresh essential oils will burn to trigger feelings of femininity and beauty. Visual merchandising, decor, and store layout will work together to create a harmonious store image that is visually pleasing. Sense of touch will be satisfied though the merchandise, a comfortable seating area, and special attention to the dressing rooms, which will have soft rugs that feel good on bare feet and padded benches for sitting.

Store Niche

Several factors are involved in creating a niche in the women's retail business. Our stylist service is unique to the area. Our specialty sizing gives us an edge, including our special-order service. Our store atmosphere makes for a pleasing shopping experience. Our merchandise selection will offer women a variety of dresses not currently available under one roof. Exclusivity of merchandise will satisfy the trendy customer's need to feel distinctive and fashionable. Special orders will offer an additional level of customer service. Exceptional customer service will foster a personal relationship between the customer and the store, triggering memories of a positive past shopping experience.

Trend Research

Having an in-house stylist will make it especially important to keep up-to-date on current fashion trends. Trend research will be a constant process, occurring daily. Subscriptions to trade journals, such as *Women's Wear Daily*, will bring fashion news into Bellissima on a daily basis. Observation and follow-up of upcoming trends in the Richmond and other metropolitan areas will keep local trends fresh in the conservative Richmond market. Internet research on brands, stores, and online fashion sources can be accessed at any time during the day. Three times a year, market weeks and/or trade shows will be attended in New York. By attending these events, we will have an opportunity to network with other boutique owners, pick up new lines to keep Bellissima's merchandise selection fresh, and observe fashion trends and how they are worn. After picking up on trends, we will expose the trends to the Richmond customer through visuals, displays, and styling advice.

Bellissima's location, 5710 Patterson Avenue, is in a small strip mall at the northeast corner of the Libbie Avenue and Patterson Avenue intersection. Other tenants in the strip mall include restaurants, a consignment shop, a salon, a furniture refinisher, a home lighting store, an embroidery shop, and a printer. On the northwest corner of the intersection is a special-education school. On the southwest corner is another strip mall, consisting of a salon, a competing women's boutique, a gift shop, a toy store, and a small grocery store. On the southeast corner, directly across from our location, are two banks, a dry cleaner, and a gas station.

One of the main reasons we chose the location at 5710 Patterson Avenue was because of the proximity to other retail locations. In the area of Libbie and Grove and Libbie and Patterson, nine other women's clothing boutiques already exist, some of which are competitors. These stores, including numerous gift shops, home stores, and restaurants, give the area enough of a retail presence to make it a destination shopping location. Opening a new women's boutique in this area will fit into the existing setting and bring in shoppers who are already in the area. In addition, a popular gourmet sandwich eatery is two doors down from 5710; many potential customers who go to the restaurant will pass by our new store, giving them a chance to stop in and check out our merchandise.

It is necessary that our store is in this affluent area because the average price of our dresses is approximately $240; therefore, our customer needs to have enough discretionary income to be able to purchase dresses on a regular basis. The previous tenant of the store was a retailer as well; therefore, we have the advantage of using her leftover fixtures and incorporating them into our ideas to save money and create a new look.

Bellissima's location at 5710 Patterson is easily accessible by either bus or car. A bus stop is located at the end of the strip mall. Driving to the area is simple; there is ample street parking in front and a small parking lot in the rear. The lot is well lit at night, ensuring safety after dark for both employees and customers. In the front, streetlamps light the street at night, as well as front-of-store lighting at the entry of each retail location. Since the other businesses, excluding the restaurants, close at six o'clock each evening, we will also close at that time. Because few patrons will be in the area after six, we will not have much traffic past this hour, therefore making it unprofitable to be open.

The large window in the front of Bellissima will contain three mannequins that will model our latest inventory. Customers who are just walking by the area or who are window shopping will be enticed to come inside of our store and see what other beautiful merchandise Bellissima carries in its 800-square-foot space.

In the middle of the store's selling floor will be two four-ways displaying our newest merchandise. Two more mannequins, dressed in new merchandise, will be located further back in the line of sight of the entering customer, pulling the customer's eye farther into the store. We want the customers to view everything we have to offer; therefore, drawing them all the way into the store will benefit us greatly. The rest of our dresses will be housed in the classy built-in wooden cabinets that were left behind by the previous tenant. These cabinets are along the side walls of the store; they will hold older side-hanging merchandise and feature new merchandise on two faceouts. On the left side of the store, built-in glass shelving separates two wooden cabinets. This shelving will showcase some of our handbags and jewelry. On the right side of the store, the wooden cabinets will line the wall followed by a single glass shelved cabinet to hold special accessories. This leads into the cash/wrap area with a glass top and mirrored bottom to display the rest of our accessory items. On top of the cash/wrap we will have a jewelry display of our moderately priced items in order to stimulate a last-minute, unplanned purchase. Next to the cash/wrap on the left is another glass shelf to hold purses. The four-ways in the middle of the store will have the newest merchandise because that is what the customer will see first. The older merchandise will be displayed in less prominent areas along the sides of the store. Sale items will be located in the rearmost cabinet, near the dressing rooms and the back of the selling space.

As with the jewelry, we will only display a few purses per style so that the customer feels as though the items are more exclusive. Extra inventory will be stored in the stock room.

In the back portion of the store, which is 600 square feet, there is a storage room, a bathroom, an office, and three dressing rooms. Each dressing room has a mirror, a stool, a fuzzy carpet, and hooks to hang garments. Outside the dressing room area are more mirrors so the customer can see the garment in a different light when coming out to show her shopping companion. Having the mirrors outside the dressing room gives sales associates the opportunity to get involved and make suggestions for accessories to accompany the outfit.

In the very front of our store will be two lounge chairs and a small table for shopping companions who feel like sitting to relax. This area is important so that our main customer does not feel rushed when she is shopping because she will know that her shopping companion is being entertained. The table is also a good place for us to put a sample look book to promote our stylist services. The coffee table will have a glass top that enables all to see an extra display of jewelry in the drawer below. The location of the seating area is perfect because it is not in the customer's line of sight when entering the store; therefore, it does not take away from the merchandise fixtures and displays.

The main colors of Bellissima will be apple green, sunny yellow, and white, based on the store's logo. The interior walls will be painted yellow, to enhance the limited natural light, with a secondary color of white. The apple green will be used as an accent color. This color scheme will complement the gray carpet and white painted wooden cabinets. This will create a warm but fresh feel to the store that will make customers comfortable and happy. We will keep the new gray wall-to-wall carpet that the previous tenant had installed, but we will add some small sisal accent rugs. The bathroom will be painted apple green, with accents of white and yellow. We will replace the existing worn ceramic tile in favor of a wooden floor. Taking inspiration from the daisy from Bellissima's logo, the store will have a subtle garden theme, which will be reflected in a weekly selection of fresh flowers.

By having the majority of our inventory along the walls, Bellissima will be a more open space so the customer never feels overwhelmed or cluttered. Whether customers have made a purchase in our store or not, we want them to leave feeling relaxed, refreshed, and beautiful.

Target Customer

After talking to stores in the area, and based on Kate's familiarity with the customer base, we ascertained that the customers who would shop Bellissima would come from the following zip codes: 23226, 23221, 23229, and 23220. The women from these zip codes, aged between 13 to 65, are already the most frequent shoppers of the Grove/Libbie and Patterson/Libbie areas. The majority of shoppers are between the ages of 18 to 45; this group will be our target market. The targeted shopper from this area is financially well off, with an income above the average. The low end of the age group is made up of girls attending St. Catherine's School or the University of Richmond; both are private schools in close proximity to our store. The women aged 18 to 25, although younger, will shop in the area because either their families are financially well off and will provide a fair amount of spending money for clothing, food, and incidentals or, after graduation, because they are accustomed to shopping in the area. The rest of the women, aged 25 to 45, are either working or staying at home with children. The working women will shop the store not because of its close proximity to work but because it is located in close proximity to their homes.Bellissima's target customer is more fashion-conscious than the average woman, selecting trendy garments in addition to wardrobe staples. Since such customers select trendier items, these women will shop more frequently than a consumer who prefers classic pieces, in order to keep up with current trends. These women are early adopters or early majority adopters in fashion trends. They desire exclusive, well-made pieces that will offer them a unique look that differentiates them from other women.

After analyzing data from the U.S. Census Bureau, we ascertain that there are 12,501 women in Bellissima's target group from our selected zip codes. Data factored into the target market consisted of women between the ages of 18 to 45, college educated, and with an income of over $50,000 per year. Individual analysis of each zip code is located in Appendix A. In *Women's Wear Daily (WWD)* Atlanta supplement, dresses are highlighted as a key trend for Fall 2007 (Kleinman, 6). Many looks, from wrap dresses to trapeze looks to jersey minidresses, will carry over from Spring 2007. Highlighted is the versatility of dresses, with *WWD* noting that many consumers will layer simple day dresses over jeans or leggings in order to carry the dress into the colder months. WWD also notes that Hollywood has a major impact on the dress trend—Sienna Miller's role as Edie Sedgwick in the movie *Factory Girl* has pushed the sixties mini shift dress into the fashion "must have" list. The recent popularity of dresses will immediately bring customers to our store; retaining customers will be accomplished through marketing our store as a place known for its wide selection of day-to-evening options. We feel that the market for dresses is steady; we chose this category not on the trendiness of dresses but because we feel that the Libbie/Patterson shopper will desire a store that carries a wider range of dress types than currently available. After surveying consumers that shop in the area, it was apparent that these customers were looking for a dress store that carried more than just cocktail dresses. We feel that we can fill this demand because of our variety of casual, semicasual, and cocktail dresses that our store will carry. We also feel that the market for dresses is steady because many of our dresses will be worn to events such as weddings, bridal or baby showers, work, or other special occasions. We do not see a decline in these events. Potential growth for Bellissima would include another location in another city. The Richmond market already has a bounty of special-occasion stores, and having another branch of our store in Richmond would not only take away from the boutique feel of the store but also be unnecessary because of Richmond's relatively small size. After consumer analysis by zip code, there are not enough potential customers to shop at another location. We also do not feel that adding another product category, such as prom dresses, would be feasible because it would take away from the image of the store. We want our customer to associate our dress shop with casual as well as cocktail dresses; there is no other dress category that would complement our selection. The only other category that could possibly complement Bellissima's current selection would be skirts and blouses. These items would work with our current selection of merchandise, but would require a larger selling space because customers like a variety of tops/blouses to go with their skirt choices. This is not possible in the chosen location because of size restraints.

Pricing and Sales Strategies

Markup

Our overall store markup will be 55 percent. Our dresses will range in price from $125 to $350. This results in an average dress price of $240. We will mark up our dresses approximately 55 percent, giving them a cost of approximately $106 each, and a gross profit of $132. Packaging costs will be approximately $2, consisting of logo garment bags.

Our handbags will range in price from $50 to $200. This results in the average handbag costing $125. We will offer handbags that complement the stock of dresses, ranging from evening bags, clutches, to everyday handbags. For every dress sold, we will aim to sell a coordinating handbag, estimating success one in three times. We will mark up our handbags approximately 55 percent, giving them a cost of approximately $54 each and a gross profit of $69. Packaging costs will be approximately $2, consisting of logo shopping bags, tissue, and logo stickers.

Our jewelry will range in price from $25 to $125, resulting in the average jewelry piece costing $75. We will mark up our jewelry approximately 55 percent, giving them a cost of approximately $32 each and a gross profit of $41. Packaging costs will be approximately $2, consisting of logo shopping bags, tissue, small boxes, and logo stickers. Sale jewelry will be priced at an average of 35 percent off, resulting in an average sale price of $45.

Markdowns

Our markdowns will begin at 20 percent off and move in increments of 15 percent, until the merchandise is 75 percent off, our cutoff markdown. The markdown schedule will be as follows: 20 percent, 35 percent, 50 percent, 65 percent, 75 percent off. Store promotions will feature merchandise that is 20 percent off, as well as being the start of the markdowns for any given product. We will put more emphasis on 35 percent and 50 percent off, allowing the store to keep a small amount of profit. We will leave the 65 percent and 75 percent off markdowns for end-of-season clearances. The remaining items that are not sold at 75 percent off will then be sold to a liquidator.

Sale dresses will be priced at an average of 35 percent off, resulting in an average sale price of $156. We estimate that only 40 percent of total dresses will be sold on sale, as we will coordinate our vendor shipments to complement peak seasonal dress purchases.

We estimate that only 25 percent of total jewelry will be sold on sale. The jewelry on the lower end of the price range ($25 to $50) will be modestly priced, making it easy for customers to justify an additional small purchase. The majority of our jewelry will be in the price range of $25 to $50, so it has a higher turnover rate. The jewelry that will be placed on sale will only be the more expensive pieces, ranging from $50 to $125. Pieces in this range are harder for the customer to justify purchasing in order to wear only once or twice.

Sale handbags will be priced at an average of 35 percent off, resulting in an average sale price of $81.We estimate that only 40 percent of total handbags will be sold on sale, since the fashion cycle for evening bags and clutches does not change rapidly.

Multiple Pricing

Our store will not offer multiple pricing.

Competition's Pricing

Our direct competitors have similar pricing schedules as Bellissima. All of these stores are boutiques catering to the same target group; the success of these businesses outlines the fact that the Libbie/Grove customer is willing to pay more than the average customer.

Our indirect competitors vary from Bellissima in pricing. Neiman Marcus, Saks Fifth Avenue, and Nordstrom are department stores that offer a wider selection of merchandise than Bellissima. Since these stores also cater to a wider variety of customers, they are able to offer merchandise that is higher priced, lower priced, and comparable to Bellissima. White House Black Market has a pricing schedule that is on the low side compared to Bellissima. Their dresses cost between $128 to $198, with the average price being $148. Pink has a pricing schedule that is on the high side compared to Bellissima. Their average dress price is $450.

Stylist Service Pricing

Fee schedule for our in-house stylist is as follows:

- In-store stylist help while shopping: Complimentary, including Polaroid photos of possible outfits selected.
- Out-of-store stylist help while shopping at other stores: $50 per hour for first hour; $35 for all subsequent hours, including Polaroid photos of possible outfits selected.
- In-home consultation: $120 for a two-hour consultation, $50 per hour for all subsequent hours. Outfit consultation includes a look book consisting of at least 30 different complete outfits; closet consultation, not including costs of new hangers and/or storage boxes, includes sorting through existing clothing and organizing customer's closet.

Customer Service

We will accept Visa, MasterCard, American Express, personal checks, and cash. Through Merchant Services, a credit card processing company, Visa, MasterCard, and American Express will be processed. Visa and MasterCard charge 0.05 percent of each transaction, American Express charges 0.15 percent of each transaction. We will not get TeleCheck service to verify personal checks, as the low volume of checks would not offset the cost of the service.

Being a boutique store, we will be able to distinguish ourselves from bigger stores through exceptional customer service—going above and beyond what large, mass retailers can offer. Upon entering the store, customers will be greeted and offered a beverage of either fresh-flavored mint or lemon water, sparkling water, or white wine if age-appropriate. Children will have a selection of toys and coloring books to occupy them. A cozy seating area, near the dressing rooms, will provide a variety of magazines and a comfortable place to sit for those accompanying the shopper. A sales associate will then individually work with the customer to put together an appropriate outfit while offering fashion advice and showing the customer new merchandise.

Each dress will be offered in four sizes, in order to create exclusivity and assure the shopper that her outfit will be unique at any social function. Our sizes will range from 0 to 8, with the offer of special order for sizes not offered in the store. For those that spend over $500 on their visit, the associate who assisted them will send a handwritten thank-you note to let the customer know how much we appreciate their business. By writing notes, we aim to build on customer loyalty and create a personal relationship between the customer and the store.

Competition

Direct Competitors

Our direct competitors are Levy's, A.R. Bevans, J. McLaughlin, Monkey's, and D.M. Williams. These are all stores that are in the Libbie/Grove or Libbie/Patterson shopping area. These stores are all boutiques in nature and vary only slightly in their merchandise selection, sometimes even overlapping in actual merchandise. We have included these five stores as our direct competition because a shopper goes to the Libbie Avenue area as a shopping destination and would typically visit these five shops on each visit. Since the Libbie shopping area is known to contain mid- to high-priced boutiques, people often visit the area in order to find special-occasion wear or an out-of-the-ordinary dress that can take them from work to evening.

Levy's, a Richmond institution for many years, caters to middle-aged women by offering suiting pieces, fun everyday wear, and a small selection of special occasion. It can be noted that the special-occasion wear on sale at Levy's is quite stale; by not marking down special-occasion pieces, there is no clearing out of old merchandise, some of which has been around for more than two years. Levy's does have a devoted following and attentive customer service. Levy's also has a sister store in Charlottesville, Virginia.

A.R. Bevans is targeted to a hipper, trendier crowd. Most of the store's selection is based on fun, eye-catching everyday wear, with suiting, denim, and special occasion offered in small amounts. A.R. Bevans offers special occasion wear only during the holiday season of October through December; few casual dresses are available in the store. The major drawback to A.R. Bevans is an extremely strict return/exchange policy that makes it difficult for indecisive customers to purchase at the store.

J. McLaughlin is a small privately owned chain that has 36 stores in 15 mostly northeastern states. Under its own private label, J. McLaughlin carries both men's and women's clothing and accessories targeted to a middle-aged, conservative consumer. The store is well merchandised and very pleasant to be in, but the merchandise offerings are not very extensive. Most of the offerings are the same item, differing only in color or pattern.

Monkey's is a store that specializes in special-occasion merchandise. They target a middle-aged consumer and carry the widest variety of price points in the area: from $200 dresses on up to custom orders that can top $1,500 or more. Monkey's selection is thin; this is due in part to a recent change in owners. The new owners are trying to reposition Monkey's image to a younger crowd and move away from solely special-occasion wear.

D.M. Williams is not a competitor in terms of product selection but is still included because they do carry the occasional dress mixed in with their work wear pieces. Shoppers walking around the Libbie area would stop into D.M. Williams and might come across a dressier piece, especially during the holiday season. D.M. Williams has another location in Virginia Beach, which means that the store could always transfer an item from one store to another. They target the working woman.

Indirect Competitors

Our indirect competitors consist of Saks Fifth Avenue, Nordstrom, Neiman Marcus, White House Black Market, and Pink. These competitors communicate to the customer on many different media levels. The large chain stores have a lot more money and resources available to promote and communicate their product and message to our target customer. Catalogs are one way that our competitors communicate with customers. These catalogs allow customers to view the new products and also inform the customer of any promotions that are occurring. Department stores have large seasonal sales to clear out old merchandise or have the ability to move old merchandise to their other stores such as Saks Off Fifth Avenue and Nordstrom Rack. The chain department stores notify customers through coupons in the mail regarding their sales and by placing advertisements in high-profile newspapers and magazines. White House Black Market has a frequent-buyers program that gives consumers incentives to shop by giving them five percent off every purchase. Other large chains offer credit cards as well.

When customers sign up for these programs, it gives the store many different forms of contact information to use in the future to help build relationships with a customer, such as their e-mail address and home address. This information also gives stores new ways to contact customers to inform them of sales and special promotions. In addition, these cards give stores a way to track consumer's purchases and use that information to determine what products are the most popular, helping them determine what product assortment they should carry. Specialty boutiques like Pink in Carytown will call and send personal thank-you cards to previous clients. This type of communication helps build personal relationships with the customers to gain customer loyalty and repeat purchases in the future. When the relationship is formed, employees can cater their customer service to that particular customer because they already know their likes and dislikes.

The customer service of our competitors is very good. Nordstrom is known for their incredible return/exchange policy where they offer refunds to customers for virtually anything. They also operate under the pretense that the customer is always right. The large chain stores can provide the service of calling their other stores in order to find the right color or size that the customer wants. White House Black Market will search the entire country in order to satisfy their customer needs. Boutiques like Pink can communicate with the manufacturer or company to special-order an item if it is not in stock in the correct size the customer needs. The small boutiques can also give intimate and specialized customer service. This customer service helps cater the shopping experience specifically to the individual and ultimately leads to a higher potential of a sale.

Our indirect competitors have much strength with regard to communicating to the customers and have well-established names in the fashion world; conversely, some of their weaknesses are our strengths. The competitors cannot guarantee the service that our store will provide. Especially in large department stores, it is more difficult to cater to specific customer needs and develop intimate relationships with customers. Large department stores can also be disorganized, crowded, and not appear as neat and organized as our small boutique. This is especially true during the holiday season and sales events. These larger stores cannot give the customer the feeling of exclusivity with their product when they carry large amounts of the same dress in the same size and have many stores. In our store we will be able to evoke feelings of individuality and "one of a kind" because there are such a limited number of items in stock. Many times our customer will be buying the only garment in that color, style, and size that is on display. This is never the case in the larger stores. Exclusivity is a big perk because our clientele does not want to have the same dress as anyone else, especially for a special event. Unlike boutiques, department stores have the advantage of a wider selection in clothing and price points, which is important to the customer who wants many options.

Advertising and Promotion

Communication to Customers

Located on the corner of Libbie and Patterson Avenues, Bellissima offers a variety of exclusive dresses, handbags, and jewelry for trendy women in a comfortable setting. Exceptional customer service and a personal stylist will make finding the perfect dress for any occasion fun and easy, whether in-house or around town.

Methods

We will use a variety of methods to keep our customers informed about our business. Print ads through *Richmond Magazine*, *Richmond Bride* (a supplement to *Richmond Magazine*), and *Style Weekly* will complement informational postcards sent to current customers on our mailing list. *Richmond Magazine* has been created specifically for residents of Richmond who have purchased this magazine, making them an attentive audience to the advertisements. *Style Weekly* has a large reach of Richmond residents.

Previous to the store opening, we will feature a quarter-page color advertisement in the September issue of *Richmond Magazine*, at a cost of $2,100. The purpose of this ad will be to create awareness of our store and its products and services. In September, December, March, and June, *Style Weekly* puts out *Cue*, a quarterly fashion supplement. In *Cue*, we will continue to advertise the store's opening and the products and services we offer. In the September *Cue* ad, we will also advertise our grand-opening cocktail party. This party will be held on a Thursday evening from 6:00 p.m. to 8:00 p.m., to raise awareness. A door prize will be given, and customers will have a chance to bring a friend along to view our merchandise, shop, and become familiar with the staff. Customers will also have a chance to sign up for our mailing list and receive a voucher for 15 percent off any one item. We will run an additional ad in the March issue of *Cue* in order to bolster the sales on our new spring merchandise. Postcards will be mailed to those who signed up to be on our mailing list informing them of special events and information—for example, the arrival of a new designer or a sale.

In addition to these forms of print advertisements, we will also have a Web site for our customers to find out more information about our store. This Web site will include information about our location, contact information, information about our staff, the designers we carry, and information about our special stylist services we provide. We will have many visuals on the Web site to show customers our store image and the type of merchandise we offer in order to lure them in to visit our store. Due to financial limitations, we will create the Web site ourselves using a free online Web site creator such as www.ebizwebpages.com/main/. Once it has been created, it will only cost $20 per month to keep the Web site up and running.

Background

Kate Aliberti and Samantha Baum have a combined expertise that will enable them to successfully open and manage a small boutique. Both Kate, in her current position as store manager at Anthropologie, and Samantha, in her past position as store manager at Arden B, excel in managing all aspects of store operational controls. They also have been responsible for overseeing the hiring and training of all new employees in their respective stores, with Samantha being recognized for decreasing Arden B's turnover rate. As store managers, both Kate and Samantha are familiar with utilizing weekly sales reports to track, analyze, and communicate business results and determine strategies to maximize sales.

In her current position as women's accessories buyer at Bloomingdale's, Samantha has learned the complete process for making a successful six-month buying plan. She also has familiarity with vendor relations and keeps current on current fashion trends.

In Kate's past position as manager at A.R. Bevans, she intimately knows the Libbie/Grove and Libbie/Patterson consumer. This personal knowledge of the store's target market will help the business cater to the desires of the clientele and ensure a specialized shopping experience. Refer to Appendix C for more background details.

Strengths and Weaknesses

Whereas Samantha's strengths lie in the behind-the-scenes operations, Kate excels at front-of-the-house activities. Therefore, Samantha's and Kate's respective skill bases complement each other. Samantha prefers math and financials. She is organized and already knows the details of a successful buying plan. Consequently, Samantha will be in charge of store financial statements, working closely with the bookkeeper to analyze sales plans and payroll expenses.

Kate is more of a people person. She thrives on personal interaction and likes to be out on the sales floor getting to know her customers. Kate will be in charge of hiring and training new employees, scheduling, dealing with customer issues, and selling merchandise. In dividing the workload into front-of-the-house and back-of-the-house responsibilities, both Kate and Samantha will be responsible for different aspects of the store, enabling them to devote their full energy into running a profitable retail enterprise.

Compensation

During the first year in business, Bellissima will be run with two owner/managers, Samantha and Kate, and one part-time employee. Samantha and Kate, working an average of 40 hours per week each, will make $400 per week or about $21,000 each the first year. The part-time employee, working 15 hours per week, will make $150 per week, or about $7,900 per year. No benefits or paid vacation time will be offered in the first year. Hopefully, these perks, including a pay raise, will be offered to the owner/managers as the store begins to generate more profit.

Samantha and Kate, the owner/managers, will both work five days a week. Both will be in the store from Tuesday to Friday, and they will alternate Mondays and Saturdays as days off in order to have a two-day weekend. On these days off, the part-time employee will assist in order for two people to be working at the store at all times.

SALES PLAN EXERCISE	
Customers per day	6
Average purchase	$140
Daily sales	$850
Days open annually	307
Annual sales	$260,950

SALES REVENUE FORECAST		
MONTH	**SALES, IN DOLLARS**	**SALES, IN PERCENT**
September	$20,660	7.9%
October	$19,570	7.5%
November	$21,750	8.3%
December	$26,100	10.0%
January	$17,400	6.7%
February	$21,740	8.3%
March	$23,920	9.2%
April	$23,920	9.2%
May	$23,920	9.2%
June	$23,920	9.2%
July	$19,570	7.5%
August	$18,480	7.1%
TOTAL	**$260,950**	**100%**

Sales per selling square foot: $326
Selling square feet: 800
Total Year 1 sales: $260,950

September $20,660 (–5%)
Slightly below average due to just opening, but bolstered by promotional efforts and the novelty of being a new store

October $19,570 (–10%)
Below average due to just opening and traditional slow month of October

continued

November	$21,750 (average) Average, due to pickup in sales after the Thanksgiving holiday and increased purchases on holiday event wear
December	$26,100 (+20%) Above average, due to holiday events and New Year's Eve
January	$17,400 (–20%) Below average, due to after-holiday decline in the retail sector and end-of-season clearances
February	$21,740 (average) Average, due to increased sales around Valentine's Day
March	$23,920 (+10%) Above average, due to increase in promotional efforts linked with new spring merchandise
April	$23,920 (+10%) Above average, due to new spring merchandise and local social events, such as horse races and spring socials
May	$23,920 (+10%) Above average, due to events linked to graduations and weddings
June	$23,920 (+10%) Above average, due to events linked to weddings and new merchandise
July	$19,570 (–10%) Slightly below average, due to summer vacations and slowdown in weddings and other social events
August	$18,480 (–15%) Below average, due to cessation of summer merchandise purchasing, end-of-season clearance, and summer vacation
Year 1 Total	**$260,950**

BREAK-EVEN SALES REVENUE FORECAST Year 1, Per Annum	
	DOLLAR
Annual Sales	$260,950
Annual Fixed Costs	$108,390
Gross Profit	45.4%
Break-Even Sales	$238,744
Sales over Breakeven	$22,206
Profit	$10,081

SALES AND GROSS MARGIN EXERCISE							
	REGULAR DRESSES	SALE DRESSES	REGULAR HANDBAGS	SALE HANDBAGS	REGULAR JEWELRY	SALE JEWELRY	
Average Cost	$106	$106	$54	$54	$32	$32	
Bags and Wrap	$2	$2	$2	$2	$2	$2	
Total Cost	$108	$108	$56	$56	$34	$34	
Average Selling Price	$240	$156 (35%)	$125	$81 (-35%)	$75	$49 (-35%)	
Gross Profit	$132	$46	$69	$25	$41	$15	
Gross Profit Percent	55.0%	29.5%	55.2%	30.9%	54.6%	30.7%	
Annual Sales Percent	42%	28%	9%	6%	11%	4%	
Annual Sales	$109,599	$73,066	$23,486	$15,657	$28,705	$10,438	$260,950
Annual Gross Profit	$60,279	$21,554	$12,964	$4,838	$15,673	$3,204	$118,513
Total Gross Profit	45.4%	= ($118,513/$260,950)					

Assumptions:
Total sales percent: 70% Dresses, 15% Handbags, 15% Jewelry
Dress Sales: 60% Full price, 40% Sale price
Handbag Sales: 60% Full price, 40% Sale price
Jewelry Sales: 75% Full price, 25% Sale price

Sales and Gross Margin Justification

Bellissima will offer a variety of dresses, handbags, and jewelry for the trendy women aged 18 to 45. Our store will give customers a variety of types of dresses: cocktail, special-occasion, and party dresses; day/sun dresses; and work-appropriate dresses. After reviewing competitor stores (Levy's, A.R. Bevans, D.M. Williams, J. McLaughlin, Monkey's) and finding a lack of variety in dress type, we feel that this range of dresses will be a welcome addition to the shopping area. This variety, along with an optional in-house stylist, will give our store distinctiveness not yet available in the Libbie and Grove area.

First-year sales will be $260,950. Based on the 800 square feet of selling space that our store provides, the average sales per square foot will be $326. We will aim to have a daily sales goal of $1,000 (rounded from $1,013) by the third year, similar to main competitor A.R. Bevans. In the

continued

first year, though, we will have a more obtainable daily sales goal of $850, based on the selling price of two full-price dresses, one sale dress, one full-price handbag, and two sale jewelry pieces ($240 + $240 + $156 + $125 + $49 + 49 = $859). Like surrounding competitors, Bellissima will be open 307 days a year. This means that we will hold store hours from 10 to 6 Monday through Saturday. We will be closed on Sundays, New Year's Day, Memorial Day, Fourth of July, Christmas, Thanksgiving Day, and Labor Day.

For our overall sales, 70 percent will be dresses (about three dresses a day), 15 percent will be handbags (about one a day), and 15 percent will be jewelry (about two pieces a day). The storewide total gross profit will be 45.4 percent, based on total gross profit dollars of $118,513.

Our markdowns will begin at 20 percent off and move in increments of 15 percent, until the merchandise is 75 percent off, our cutoff markdown. The markdown schedule will be as follows: 20 percent, 35 percent, 50 percent, 65 percent, 75 percent off. Store promotions will feature merchandise that is 20 percent off, as well as being the start of the markdowns for any given product. We will put more emphasis on 35 percent and 50 percent off, allowing the store to keep a small amount of profit. We will leave the 65 percent and 75 percent off markdowns for end-of-season clearances. The remaining items that are not sold at 75 percent off will then be sold to a liquidator.

Dresses

Bellissima's dresses will range in price from $175 to $350. This results in an average dress price of $240. We will mark up our dresses approximately 55 percent, giving them a cost of approximately $106 each and a gross profit of $132. Packaging costs will be approximately $2, consisting of logo garment bags. Sale dresses will be priced at an average of 35 percent off, resulting in an average sale price of $156. We estimate that only 40 percent of total dresses will be sold on sale, as we will coordinate our vendor shipments to complement peak seasonal dress purchases.

Handbags

Bellissima's handbags will range in price from $50 to $200. This results in the average handbag costing $125. We will offer handbags that complement the stock of dresses ranging from evening bags, clutches, to everyday handbags. For every dress sold, we will aim to sell a coordinating handbag, estimating success one in three times. We will mark up our handbags approximately 55 percent, giving them a cost of approximately $54 each and a gross profit of $69. Packaging costs will be approximately $2, consisting of logo shopping bags, tissue, and logo stickers. Sale handbags will be priced at an average of 35 percent off, resulting in an average sale price of $81. We estimate that only 40 percent of total handbags will be sold on sale, since the fashion cycle for evening bags and clutches does not change rapidly.

Jewelry

Bellissima's jewelry will range in price from $25 to $125. This results in the average jewelry piece costing $75. We will offer costume and semiprecious earrings, bracelets, rings, and necklaces that will complement dresses in stock. For every dress sold, we will aim to sell coordinating jewelry, estimating success two in three times. We will mark up our jewelry approximately 55 percent, giving them a cost of approximately $32 each, and a gross profit of $41. Packaging costs will be approximately $2, consisting of logo shopping bags, tissue, small boxes, and logo stickers. Sale jewelry will be priced at an average of 35 percent off, resulting in an average sale price of $45. We estimate that only 25 percent of total jewelry will be sold on sale, since the jewelry on the lower end of the price range ($25 to $50) will be modestly priced; therefore, it will be easy for customers to justify an additional small purchase. The majority of our jewelry will be in the price range of $25 to $50, as it has a higher turnover rate. The jewelry that will be placed on sale will only be the more expensive pieces, ranging from $50 to 125. Pieces in this range are harder for the customer to justify purchasing in order to wear only once or twice.

FIXED COSTS FORECAST Per Annum	
Accounting	$3,900
Alarm Services	$1,200
Banking Services	$1,800
Electric	$2,100
Gas and Water	$2,400
Insurance	$5,400
Maintenance	$2,400
Miscellaneous	$2,400
POS System Maint.	$1,200
Rent	$20,964
Staffing	$49,404
Staffing Taxes	$6,912
Supplies	$1,800
Telephone/Internet	$1,680
Travel	$4,830
TOTAL	**$108,390**

CAPITAL SPENDING PLAN	
CAPITAL ITEMS	
Fixtures: Two chrome 4-ways, 4 rolling racks, jewelry displays, two glass shelving units envelopes, postcards	$545
Seating Area: Two armchairs, side table, lamp, jute rug, coffee table, accent pillows	$2,050
Mannequins: 6 French dress forms with base and neck block	$615
Decor: Pictures, stools in fitting rooms, potted plants, etc.	$1,000
Signage: Outdoor awning with printed logo, open/closed/sale signs, logo decal for window	$1,700
Kitchen Supplies: Microwave, mini fridge	$260
Fax Machine	$125
Printer/Copier	$175

continued

CAPITAL SPENDING PLAN *(continued)*	
CAPITAL ITEMS	
Computer: Two desktop computers, one for office, one as register	$1,500
Office Furniture: Desk, office chair, filing cabinet, shelving unit	$640
Small Office Equipment: Scissors, staplers, telephones, calculator, wastebaskets, etc.	$200
Leasehold Improvements: DIY: Paint for exterior, interior; retile bathroom and entryways	$800
Vacuum	$255
Software: Microsoft Office, Isis POS system	$180
Steamer	$800
TOTAL	**$14,545**
EXPENSE ITEMS	
Advertising Stationery: Business cards, letterhead, envelopes, postcards	$2,630 $475
Opening Inventory	$31,200
Lease Deposit: One month's rent	$1,750
Business License/Permit	$230
Opening Promotion	$3,000
Utilities/Phone Connection	$50
Supplies: Office: Pens, paper, folders, etc.	$420
Selling: Tagging gun, tags, hangers	
TOTAL	**$39,755**
TOTAL	
Total Capital Items	$14,545
Total Expense Items	$39,755
Contingency Reserve (10%)	$5,430
TOTAL	**$59,730**

Start-up Justification

Start-up Costs
The majority of Bellissima's costs are fixed and, therefore, permanent. With the few costs that are negotiable, we have controlled them to get by with as little as possible. One of the biggest expenses that we were able to control was the fixturing. The previous tenant was also a retail business and had custom built-in cabinets made to house her merchandise. These built-ins were left in the building, and we will take advantage and use them. We will also remain with the newly laid gray wall-to-wall carpeting. We will, however, give everything a fresh coat of paint and retile the bathroom floor. In order to save costs, Kate will enlist her brother, a contractor, to lay the tile for the store, at no cost.

Loan
Both Kate and Samantha will be investing $11,000 each into the business. Kate's contribution will

Forecasted Financial Statements: Years 1–3

PROFIT AND LOSS STATEMENT Year 1: September 1, 2007 to August 31, 2008						
	SEPT	**OCT**	**NOV**	**DEC**	**JAN**	**FEB**
NET SALES	$20,660	$19,570	$21,745	$26,095	$17,400	$21,745
COST OF GOODS SOLD	$11,280	$10,685	$11,873	$14,248	$9,500	$11,873
GROSS PROFIT	$9,380	$8,885	$9,872	$11,847	$7,900	$9,872
ITEMIZED EXPENSES						
Accounting	$300	$300	$300	$300	$300	$300
Advertising/Promo.	$2,630	$20	$20	$1,715	$20	$20
Alarm Services	$100	$100	$100	$100	$100	$100
Banking Services	$150	$150	$150	$150	$150	$150
Electric	$175	$175	$175	$175	$175	$175
Gas and Water	$100	$300	$300	$300	$300	$300
Insurance	$450	$450	$450	$450	$450	$450
Maintenance	$200	$200	$200	$200	$200	$200
Miscellaneous	$200	$200	$200	$200	$200	$200
POS System Maintenance	$100	$100	$100	$100	$100	$100
Rent	$1,747	$1,747	$1,747	$1,747	$1,747	$1,747
Staffing	$3,791	$3,791	$4,117	$4,117	$3,467	$3,791
Staffing Taxes	$576	$576	$576	$576	$576	$576
Supplies	$150	$150	$150	$150	$150	$150
Telephone/Internet	$140	$140	$140	$140	$140	$140
Travel	$1,460	$50	$50	$50	$50	$1,460
Total Fixed Expenses	$9,639	$8,429	$8,755	$8,755	$8,105	$9,839
TOTAL EXPENSES	**$12,269**	**$8,449**	**$8,775**	**$10,470**	**$8,125**	**$9,859**
NET PROFIT/LOSS	**($2,889)**	**$436**	**$1,097**	**$1,377**	**($225)**	**$13**

come from selling off $6,000 worth of stocks, and from a generous gift from her parents in the amount of $5,000. Samantha will be withdrawing $4,500 from her savings account, cashing in $1,500 worth of stocks, and giving up her cash gift of $5,000 from her grandmother. Both women have left enough in personal savings to get by financially until Bellissima makes enough profit to constitute a salary increase. Kate's husband will also help support her financially.

After completing the capital spending plan, it became apparent that the business would need $55,770 to cover expenses. The number was then rounded and another $1,000 bumper added on the total. This resulted in $57,000 needed to get the business up and running. Since Kate and Samantha are contributing $22,000 of their personal money, a loan in the amount of $35,000 was needed.

After looking on the SBA Web site, a loan was found through SunTrust in Richmond. Since the owners of Bellissima are both women, Kate and Samantha qualified for the Community Express Loan. This allowed Bellissima's owners to pay interest on the prime rate plus 2.25 percent. Since the prime rate in April of 2007 was 8.25 percent, this means that a total interest amount of 10.5 percent will be paid over the course of the three-year loan.

MARCH	APRIL	MAY	JUNE	JULY	AUGUST	TOTAL	
$23,920	$23,920	$23,920	$23,920	$19,570	$18,485	**$260,950**	**100.0%**
$13,060	$13,060	$13,060	$13,060	$10,685	$10,093	**$142,479**	**54.6%**
$10,860	$10,860	$10,860	$10,860	$8,885	$8,392	**$118,513**	**45.4%**
$400	$500	$300	$300	$300	$300	$3,900	1.5%
$2,630	$20	$20	$530	$20	$20	$7,665	2.9%
$100	$100	$100	$100	$100	$100	$1,200	0.5%
$150	$150	$150	$150	$150	$150	$1,800	0.7%
$175	$175	$175	$175	$175	$175	$2,100	0.8%
$300	$100	$100	$100	$100	$100	$2,400	2.0%
$450	$450	$450	$450	$450	$450	$5,400	2.1%
$200	$200	$200	$200	$200	$200	$2,400	0.9%
$200	$200	$200	$200	$200	$200	$2,400	0.9%
$100	$100	$100	$100	$100	$100	$1,200	0.5%
$1,747	$1,747	$1,747	$1,747	$1,747	$1,747	$20,964	8.0%
$3,467	$4,117	$4,117	$3,791	$3,467	$3,467	$45,500	17.4%
$576	$576	$576	$576	$576	$576	$6,912	2.6%
$150	$150	$150	$150	$150	$150	$1,800	0.7%
$140	$140	$140	$140	$140	$140	$1,680	0.6%
$50	$50	$50	$1,460	$50	$50	$4,830	1.9%
$8,205	$8,755	$8,555	$9,639	$7,905	$7,905	$104,486	40.0%
$10,835	**$8,775**	**$8,575**	**$10,169**	**$7,925**	**$7,925**	**$112,151**	**43.0%**
$25	**$2,085**	**$2,285**	**$691**	**$960**	**$467**	**$6,362**	**2.4%**

PROFIT AND LOSS STATEMENT Year 2: September 1, 2008 to August 31, 2009						
	SEPT	OCT	NOV	DEC	JAN	FEB
NET SALES	$22,430	$21,430	$22,850	$28,390	$19,350	$23,570
COST OF GOODS SOLD	$12,247	$11,701	$12,476	$15,501	$10,565	$12,869
GROSS PROFIT	$10,183	$9,729	$10,374	$12,889	$8,785	$10,701
ITEMIZED EXPENSES						
Accounting	$300	$300	$300	$300	$300	$300
Advertising/Promo.	$3,305	$20	$20	$3,815	$20	$20
Alarm Services	$100	$100	$100	$100	$100	$100
Banking Services	$150	$150	$150	$150	$150	$150
Electric	$175	$175	$175	$175	$175	$175
Gas and Water	$100	$300	$300	$300	$300	$300
Insurance	$450	$450	$450	$450	$450	$450
Maintenance	$200	$200	$200	$200	$200	$200
Miscellaneous	$200	$200	$200	$200	$200	$200
POS System Maintenance	$100	$100	$100	$100	$100	$100
Rent	$1,747	$1,747	$1,747	$1,747	$1,747	$1,747
Staffing	$3,791	$3,791	$4,117	$4,117	$3,467	$3,791
Staffing Taxes	$576	$576	$576	$576	$576	$576
Supplies	$150	$150	$150	$150	$150	$150
Telephone/Internet	$140	$140	$140	$140	$140	$140
Travel	$1,460	$50	$50	$50	$50	$1,460
Total Fixed Expenses	$9,639	$8,429	$8,755	$8,755	$8,105	$9,839
TOTAL EXPENSES	**$12,944**	**$8,449**	**$8,775**	**$12,570**	**$8,125**	**$9,859**
NET PROFIT/LOSS	**($2,761)**	**$1,280**	**$1,599**	**$319**	**$660**	**$842**

PROFIT AND LOSS STATEMENT Year 3: September 1, 2009 to August 31, 2010						
	SEPT	OCT	NOV	DEC	JAN	FEB
NET SALES	$24,605	$23,310	$25,900	$31,080	$20,720	$25,900
COST OF GOODS SOLD	$13,434	$12,727	$14,141	$16,970	$11,313	$14,141
GROSS PROFIT	$11,171	$10,583	$11,759	$14,110	$9,407	$11,759
ITEMIZED EXPENSES						
Accounting	$300	$300	$300	$300	$300	$300
Advertising/Promo.	$3,305	$20	$530	$3,815	$20	$530
Alarm Services	$100	$100	$100	$100	$100	$100
Banking Services	$150	$150	$150	$150	$150	$150
Electric	$175	$175	$175	$175	$175	$175
Gas and Water	$100	$300	$300	$300	$300	$300
Insurance	$450	$450	$450	$450	$450	$450
Maintenance	$200	$200	$200	$200	$200	$200
Miscellaneous	$200	$200	$200	$200	$200	$200

MARCH	APRIL	MAY	JUNE	JULY	AUGUST	TOTAL	
$26,125	$26,125	$26,125	$26,125	$21,29	$20,160	**$283,975**	**100.0%**
$14,264	$14,264	$14,264	$14,264	$11,627	$11,007	**$155,050**	**54.6%**
$11,861	$11,861	$11,861	$11,861	$9,668	$9,153	**$128,925**	**45.4%**
$400	$500	$300	$300	$300	$300	$3,900	1.4%
$2,630	$530	$20	$530	$20	$20	$10,950	3.9%
$100	$100	$100	$100	$100	$100	$1,200	0.4%
$150	$150	$150	$150	$150	$150	$1,800	0.6%
$175	$175	$175	$175	$175	$175	$2,100	0.7%
$300	$100	$100	$100	$100	$100	$2,400	0.8%
$450	$450	$450	$450	$450	$450	$5,400	1.9%
$200	$200	$200	$200	$200	$200	$2,400	0.8%
$200	$200	$200	$200	$200	$200	$2,400	0.8%
$100	$100	$100	$100	$100	$100	$1,200	0.4%
$1,747	$1,747	$1,747	$1,747	$1,747	$1,747	$20,964	7.4%
$3,467	$4,117	$4,117	$3,791	$3,467	$3,467	$45,500	16.0%
$576	$576	$576	$576	$576	$576	$6,912	2.4%
$150	$150	$150	$150	$150	$150	$1,800	0.6%
$140	$140	$140	$140	$140	$140	$1,680	0.6%
$50	$50	$50	$1,460	$50	$50	$4,830	1.7%
$8,205	$8,755	$8,555	$9,639	$7,905	$7,905	$104,486	36.8%
$10,835	**$9,285**	**$8,575**	**$10,169**	**$7,925**	**$7,925**	**$115,436**	**40.7%**
$1,026	**$2,576**	**$3,286**	**$1,692**	**$1,743**	**$1,228**	**$13,489**	**4.7%**

MARCH	APRIL	MAY	JUNE	JULY	AUGUST	TOTAL	
$28,490	$28,490	$28,490	$28,490	$23,310	$22,015	**$310,800**	**100.0%**
$15,556	$15,556	$15,556	$15,556	$12,727	$12,020	**$169,697**	**54.6%**
$12,934	$12,934	$12,934	$12,934	$10,583	$9,995	**$141,103**	**45.4%**
$400	$500	$300	$300	$300	$300	$3,900	1.3%
$2,630	$2,630	$20	$530	$20	$20	$14,070	4.5%
$100	$100	$100	$100	$100	$100	$1,200	0.4%
$150	$150	$150	$150	$150	$150	$1,800	0.6%
$175	$175	$175	$175	$175	$175	$2,100	0.7%
$300	$100	$100	$100	$100	$100	$2,400	0.8%
$450	$450	$450	$450	$450	$450	$5,400	1.7%
$200	$200	$200	$200	$200	$200	$2,400	0.8%
$200	$200	$200	$200	$200	$200	$2,400	0.8%

continued

PROFIT AND LOSS STATEMENT *(continued)* Year 3: September 1, 2009 to August 31, 2010						
	SEPT	OCT	NOV	DEC	JAN	FEB
POS System Maintenance	$100	$100	$100	$100	$100	$100
Rent	$1,747	$1,747	$1,747	$1,747	$1,747	$1,747
Staffing	$3,791	$3,791	$4,117	$4,117	$3,467	$3,791
Staffing Taxes	$576	$576	$576	$576	$576	$576
Supplies	$150	$150	$150	$150	$150	$150
Telephone/Internet	$140	$140	$140	$140	$140	$140
Travel	$1,460	$75	$75	$75	$75	$1,460
Total Fixed Expenses	$9,639	$8,454	$8,780	$8,780	$8,130	$9,839
TOTAL EXPENSES	**$12,944**	**$8,474**	**$9,310**	**$12,595**	**$8,150**	**$10,369**
NET PROFIT/LOSS	**($1,773)**	**$2,109**	**$2,449**	**$1,515**	**$1,257**	**$1,390**

Profit and Loss Statement Justifications
Year 1: September 1, 2007 to August 31, 2008

Accounting	Payroll and bookkeeping to be done by accountant at $300 per month; increase in billable hours in late March, early April to complete taxes for IRS.
Advertising/ Promotion	September and March will be the big advertising/promotion months for the store; during the first year, intensive advertising and promotional efforts will be held from September through December in order to drum up business during initial opening and dressing for holiday events.
Alarm Services	Through Richmond Alarm; monthly fee covers cost for alarm maintenance, panic-button services, and 24-hour alarm coverage.
Banking Services	Through Wachovia, which is located conveniently to the business; covers cost of business account including costs for cashier's checks and money orders; also included will be the costs charged by the credit card processing company for Visa, MasterCard, and American Express transactions.
Electric	Through Dominion Power, standard rates, approximate cost for 1,400-square-foot retail business lighting needs.
Gas and Water	Through City of Richmond Utilities; Gas usage goes up in winter months resulting in a higher utility bill for the months of October to March, water remains constant throughout the year.
Insurance	Through Erie Insurance; liability insurance.
Maintenance	The lease is a triple net lease, meaning that the lessee is responsible for all maintenance issues that arrive; an amount will be saved each month and placed in a business savings account for when emergency maintenance issues arrive.
Miscellaneous	Extra money for unforeseen costs.

MARCH	APRIL	MAY	JUNE	JULY	AUGUST	TOTAL	
$100	$100	$100	$100	$100	$100	$1,200	0.4%
$1,747	$1,747	$1,747	$1,747	$1,747	$1,747	$20,964	6.7%
$3,467	$4,117	$4,117	$3,791	$3,467	$3,467	$45,500	14.6%
$576	$576	$576	$576	$576	$576	$6,912	14.6%
$150	$150	$150	$150	$150	$150	$1,800	0.6%
$140	$140	$140	$140	$140	$140	$1,680	0.5%
$75	$75	$75	$1,460	$75	$75	$5,055	1.6%
$8,230	$8,780	$8,580	$9,639	$7,930	$7,930	$104,711	33.7%
$10,860	**$11,410**	**$8,600**	**$10,169**	**$7,950**	**$7,950**	**$112,151**	**38.2%**
$2,074	**$1,524**	**$4,334**	**$2,765**	**$2,633**	**$2,045**	**$6,362**	**7.2%**

POS System Maintenance — Through Isis POS program; point-of-sale system maintenance will cover costs of having company serviceperson to visit store and manually work on computer and service consultation via phone; full allotted cost not used each month, extra will be placed in business savings account; serviceman charges $50 per hour for visit; $25 per hour for phone consultation.

Rent — 1,397.5 square feet x $15 per square foot.

Staffing — Two full-time employee/owners, one part-time employee.
Full-time salary: $20,800/year ($10/hour x 40 hours = $400/week);
Part-time salary: $7,800/year ($10/hour x 15 hours = $150/week).
Busy months will use all three employees, while slow months will use only the two employee/owners.

Staffing Taxes — Payroll taxes; evenly spread over 12 months, at a rate of 14 percent.

Supplies — Necessities for day-to-day business operation: stamps, copy paper, printer ink, lightbulbs, toilet paper, and so on.

Telephone/Internet — Through Verizon; quote for business rate monthly service and long distance; higher cost due to the fact that vendors are out of town and would necessitate many long-distance calls.

Travel — Travel to NYC three times a year to attend market shows: Coterie in mid-February, Accessories Circuit and Intermezzo Collection in mid-June, and Coterie in mid-September; $225 per night for four nights at discounted rate for visiting show buyers, includes cost to shuttle to and from hotel and show location, $80 per round-trip ticket from RIC to JFK on JetBlue; $50 daily allotted for food and taxis, resulting in a budget of $1460 per visit for two people; remaining purchases will be made when traveling sales rep comes to Richmond; $50 for other months will cover cost of gas for traveling requirements of stylist visiting customer homes or personal shopping expeditions for customers, increase to $75 in third year due to increase in client appointments.

CASH FLOW STATEMENT Year 1: September 1, 2007 to August 31, 2008						
	START-UP	**SEPT**	**OCT**	**NOV**	**DEC**	**JAN**
Cash on Hand (BOM)	$22,000	$9,158	$14,722	$13,516	$5,159	$4,957
Loan/Line of Credit	$40,000	$0	$0	$0	$0	$0
Cash Sales	$0	$20,660	$19,570	$21,745	$26,095	$17,400
TOTAL CASH	**$62,000**	**$29,818**	**$34,292**	**$35,261**	**$31,254**	**$22,357**
EXPENSES						
Accounting	$0	$300	$300	$300	$300	$300
Advertising/Promo.	$2,630	$2,630	$20	$20	$1,715	$20
Alarm Services	$50	$100	$100	$100	$100	$100
Amoritized Loan Pmt.	$0	$1,227	$1,227	$1,227	$1,227	$1,227
Banking Services	$0	$150	$150	$150	$150	$150
Business License	$230	$0	$0	$0	$0	$0
Computer	$1,500	$0	$0	$0	$0	$0
Decor	$1,000	$100	$100	$100	$100	$100
Electric	$20	$175	$175	$175	$175	$175
Fax Machine	$125	$0	$0	$0	$0	$0
Fixtures	$545	$0	$0	$0	$0	$0
Gas and Water	$20	$100	$300	$300	$300	$300
Insurance	$0	$450	$450	$450	$450	$450
Inventory	$31,200	$1,500	$11,000	$20,000	$14,500	$8,500
Kitchen Supplies	$260	$0	$0	$0	$0	$0
Leasehold Imprvmnt.	$800	$0	$0	$0	$0	$0
L&P Line of Credit	$0	$0	$0	$0	$0	$0
Maintenance	$0	$200	$200	$200	$200	$200
Mannequins	$615	$0	$0	$0	$0	$0
Miscellaneous	$0	$200	$200	$200	$200	$200
Office Furniture	$640	$0	$0	$0	$0	$0
POS System Maint.	$0	$100	$100	$100	$100	$100
Printer/Copier	$175	$0	$0	$0	$0	$0
Rent	$1,747	$1,747	$1,747	$1,747	$1,747	$1,747
Seating Area	$2,050	$0	$0	$0	$0	$0
Signage	$1,700	$0	$0	$0	$0	$0
Small Office Equipt.	$200	$0	$0	$0	$0	$0
Software	$4,500	$0	$0	$0	$0	$0
Staffing	$0	$3,791	$3,791	$4,117	$4,117	$3,467
Staffing Taxes	$0	$576	$576	$576	$576	$576
Stationery	$475	$0	$0	$0	$0	$0

FEB	MARCH	APRIL	MAY	JUNE	JULY	AUGUST	TOTAL
$4,405	$3,964	$7,222	$4,040	$3,558	$7,482	$9,300	
$0	$0	$0	$0	$0	$0	$0	
$21,745	$23,920	$23,920	$23,920	$23,920	$19,570	$18,485	$260,950
$26,150	**$27,884**	**$31,142**	**$27,960**	**$27,478**	**$27,052**	**$27,785**	
$300	$400	$500	$300	$300	$300	$300	$3,900
$20	$2,630	$20	$20	$530	$20	$20	$7,665
$100	$100	$100	$100	$100	$100	$100	$1,200
$1,227	$1,227	$1,227	$1,227	$1,227	$1,227	$1,227	$14,724
$150	$150	$150	$150	$150	$150	$150	$1,800
$0	$0	$0	$0	$0	$0	$0	$0
$0	$0	$0	$0	$0	$0	$0	$0
$100	$100	$100	$100	$100	$100	$100	$1,200
$175	$175	$175	$175	$175	$175	$175	$2,100
$0	$0	$0	$0	$0	$0	$0	$0
$0	$0	$0	$0	$0	$0	$0	$0
$300	$300	$100	$100	$100	$100	$100	$2,400
$450	$450	$450	$450	$450	$450	$450	$5,400
$11,000	$8,500	$17,000	$14,500	$8,500	$8,500	$11,000	$134,500
$0	$0	$0	$0	$0	$0	$0	$0
$0	$0	$0	$0	$0	$0	$0	$0
$0	$0	$0	$0	$0	$0	$0	$0
$200	$200	$200	$200	$200	$200	$200	$2,400
$0	$0	$0	$0	$0	$0	$0	$0
$200	$200	$200	$200	$200	$200	$200	$2,400
$0	$0	$0	$0	$0	$0	$0	$0
$100	$100	$100	$100	$100	$100	$100	$1,200
$0	$0	$0	$0	$0	$0	$0	$0
$1,747	$1,747	$1,747	$1,747	$1,747	$1,747	$1,747	$20,964
$0	$0	$0	$0	$0	$0	$0	$0
$0	$0	$0	$0	$0	$0	$0	$0
$0	$0	$0	$0	$0	$0	$0	$0
$0	$0	$0	$0	$0	$0	$0	$0
$3,791	$3,467	$4,117	$4,117	$3,791	$3,467	$3,467	$45,500
$576	$576	$576	$576	$576	$576	$576	$6,912
$0	$0	$0	$0	$0	$0	$475	$475

continued

CASH FLOW STATEMENT *(continued)* Year 1: September 1, 2007 to August 31, 2008						
	START-UP	**SEPT**	**OCT**	**NOV**	**DEC**	**JAN**
Steamer	$180	$0	$0	$0	$0	$0
Supplies	$440	$150	$150	$150	$150	$150
Telephone/Internet	$25	$140	$140	$140	$140	$140
Travel	$1,460	$1,460	$50	$50	$50	$50
Vacuum	$255	$0	$0	$0	$0	$0
TOTAL FIXED EXPENSES		**$9,639**	**$8,429**	**$8,755**	**$8,755**	**$8,105**
TOTAL EXPENSES	**$52,842**	**$15,096**	**$20,776**	**$30,102**	**$26,297**	**$17,952**
Cash on Hand (EOM)	$9,158	$14,722	$13,516	$5,159	$4,957	$4,405
Operating Line of Credit	$30,000	$0	$0	$1,000	$0	$0

CASH FLOW STATEMENT Year 2: September 1, 2008 to August 31, 2009					
	SEPT	**OCT**	**NOV**	**DEC**	**JAN**
Cash on Hand (EOM)	$7,058	$15,217	$15,871	$17,109	$11,602
Loan/Line of Credit	$0	$0	$0	$0	$0
Cash Sales	$22,430	$21,430	$22,850	$28,390	$19,350
TOTAL CASH	**$29,488**	**$36,647**	**$38,721**	**$45,499**	**$30,952**
EXPENSES					
Accounting	$300	$300	$300	$300	$300
Advertising/Promo.	$3,305	$20	$530	$3,815	$20
Alarm Services	$100	$100	$100	$100	$100
Amoritized Loan Pmt.	$1,227	$1,227	$1,227	$1,227	$1,227
Banking Services	$150	$150	$150	$150	$150
Business License	$0	$0	$0	$0	$0
Computer	$0	$0	$0	$0	$0
Decor	$100	$100	$100	$100	$100
Electric	$175	$175	$175	$175	$175
Fax Machine	$0	$0	$0	$0	$0
Fixtures	$0	$0	$0	$0	$0
Gas and Water	$100	$300	$300	$300	$300
Insurance	$450	$450	$450	$450	$450
Inventory	$0	$11,000	$11,000	$20,500	$16,200
Kitchen Supplies	$0	$0	$0	$0	$0
Leasehold Imprvmnt.	$0	$0	$0	$0	$0
L&P Line of Credit	$0	$0	$0	$0	$0
Maintenance	$200	$200	$200	$200	$200

FEB	MARCH	APRIL	MAY	JUNE	JULY	AUG	TOTAL
$0	$0	$0	$0	$0	$0	$0	$0
$150	$150	$150	$150	$150	$150	$150	$1,800
$140	$140	$140	$140	$140	$140	$140	$1,680
$1,460	$50	$50	$50	$1,460	$50	$50	$4,830
$0	$0	$0	$0	$0	$0	$0	$0
$9,839	**$8,205**	**$8,755**	**$8,555**	**$9,639**	**$7,905**	**$7,905**	**$104,486**
$22,186	**$20,662**	**$27,102**	**$24,402**	**$19,996**	**$17,752**	**$20,727**	**$263,050**
$3,964	$7,222	$4,040	$3,558	$7,482	$9,300	$7,058	
$0	$0	$0	$0	$0	$0	$0	

FEB	MARCH	APRIL	MAY	JUNE	JULY	AUG	TOTAL
$5,300	$7,674	$9,837	$12,750	$11,973	$12,102	$11,645	$4,759
$0	$0	$0	$0	$0	$0	$0	
$23,570	$26,125	$26,125	$26,125	$26,125	$21,295	$20,160	$283,975
$28,870	**$33,799**	**$35,962**	**$38,875**	**$38,098**	**$33,397**	**$31,805**	
$300	$400	$500	$300	$300	$300	$300	$3,900
$530	$2,630	$2,630	$20	$530	$20	$20	$14,070
$100	$100	$100	$100	$100	$100	$100	$1,200
$1,227	$1,227	$1,227	$1,227	$1,227	$1,227	$1,227	$14,724
$150	$150	$150	$150	$150	$150	$150	$1,800
$0	$0	$0	$0	$0	$0	$0	$0
$0	$0	$0	$0	$0	$0	$0	$0
$100	$100	$100	$100	$100	$100	$100	$1,200
$175	$175	$175	$175	$175	$175	$175	$2,100
$0	$0	$0	$0	$0	$0	$0	$0
$0	$0	$0	$0	$0	$0	$0	$0
$300	$300	$100	$100	$100	$100	$100	$2,400
$450	$450	$450	$450	$450	$450	$450	$5,400
$9,500	$11,800	$10,500	$17,000	$14,500	$12,500	$9,500	$143,500
$0	$0	$0	$0	$0	$0	$0	$0
$0	$0	$0	$0	$0	$0	$0	$0
$0	$0	$0	$0	$0	$0	$0	$0
$200	$200	$200	$200	$200	$200	$200	$2,400

continued

CASH FLOW STATEMENT *(continued)* Year 2: September 1, 2008 to August 31, 2009					
	SEPT	OCT	NOV	DEC	JAN
Mannequins	$0	$0	$0	$0	$0
Miscellaneous	$200	$200	$200	$200	$200
Office Furniture	$0	$0	$0	$0	$0
POS System Maint.	$100	$100	$100	$100	$100
Printer/Copier	$0	$0	$0	$0	$0
Rent	$1,747	$1,747	$1,747	$1,747	$1,747
Seating Area	$0	$0	$0	$0	$0
Signage	$0	$0	$0	$0	$0
Small Office Equipt.	$0	$0	$0	$0	$0
Software	$0	$0	$0	$0	$0
Staffing	$3,791	$3,791	$4,117	$4,117	$3,467
Staffing Taxes	$576	$576	$576	$576	$576
Stationery	$0	$0	$0	$0	$0
Steamer	$0	$0	$0	$0	$0
Supplies	$150	$150	$150	$150	$150
Telephone/Internet	$140	$140	$140	$140	$140
Travel	$1,460	$50	$50	$50	$50
Vacuum	$0	$0	$0	$0	$0
TOTAL FIXED EXPENSES	**$9,639**	**$8,429**	**$8,755**	**$8,755**	**$8,105**
TOTAL EXPENSES	**$14,271**	**$20,776**	**$21,612**	**$33,897**	**$25,652**
Cash on Hand (EOM)	$15,217	$15,871	$17,109	$11,602	$5,300
Operating Line of Credit	$0	$0	$0	$0	$0

CASH FLOW STATEMENT Year 3: September 1, 2009 to August 31, 2010					
	SEPT	OCT	NOV	DEC	JAN
Cash on Hand (EOM)	$12,578	$11,912	$13,421	$16,684	$11,842
Loan/Line of Credit	$0	$0	$0	$0	$0
Cash Sales	$24,605	$23,310	$25,900	$31,080	$20,720
TOTAL CASH	**$37,183**	**$35,222**	**$39,321**	**$47,764**	**$32,562**
EXPENSES					
Accounting	$300	$300	$300	$300	$300
Advertising/Promo.	$3,305	$20	$530	$3,815	$20
Alarm Services	$100	$100	$100	$100	$100
Amoritized Loan Pmt.	$1,227	$1,227	$1,227	$1,227	$1,227
Banking Services	$150	$150	$150	$150	$150

	FEB	MARCH	APRIL	MAY	JUNE	JULY	AUG	TOTAL
	$0	$0	$0	$0	$0	$0	$0	$0
	$200	$200	$200	$200	$200	$200	$200	$2,400
	$0	$0	$0	$0	$0	$0	$0	$0
	$100	$100	$100	$100	$100	$100	$100	$1,200
	$0	$0	$0	$0	$0	$0	$0	$0
	$1,747	$1,747	$1,747	$1,747	$1,747	$1,747	$1,747	$20,964
	$0	$0	$0	$0	$0	$0	$0	$0
	$0	$0	$0	$0	$0	$0	$0	$0
	$0	$0	$0	$0	$0	$0	$0	$0
	$0	$0	$0	$0	$0	$0	$0	$0
	$3,791	$3,467	$4,117	$4,117	$3,791	$3,467	$3,467	$45,500
	$576	$576	$576	$576	$576	$576	$576	$6,912
	$0	$0	$0	$0	$0	$0	$475	$475
	$0	$0	$0	$0	$0	$0	$0	$0
	$150	$150	$150	$150	$150	$150	$150	$1,800
	$140	$140	$140	$140	$140	$140	$140	$1,680
	$1,460	$50	$50	$50	$1,460	$50	$50	$4,830
	$0	$0	$0	$0	$0	$0	$0	$0
	$9,839	**$8,205**	**$8,755**	**$8,555**	**$9,639**	**$7,905**	**$7,905**	**$104,486**
	$21,196	**$23,962**	**$23,212**	**$26,902**	**$25,996**	**$21,752**	**$19,227**	**$278,455**
	$7,674	$9,837	$12,750	$11,973	$12,102	$11,645	$12,578	
	$0	$0	$0	$0	$0	$0	$0	

	FEB	MARCH	APRIL	MAY	JUNE	JULY	AUGUST	TOTAL
	$4,585	$8,289	$12,092	$17,347	$17,908	$19,402	$10,625	
	$0	$0	$0	$0	$0	$0	$0	
	$25,900	$28,490	$28,490	$28,490	$28,490	$23,310	$22,015	$310,800
	$30,485	**$36,779**	**$40,582**	**$45,835**	**$46,398**	**$42,712**	**$32,640**	
	$300	$400	$500	$300	$300	$300	$300	$3,900
	$530	$2,630	$2,630	$20	$530	$20	$20	$14,070
	$100	$100	$100	$100	$100	$100	$100	$1,200
	$1,227	$1,227	$1,227	$1,227	$1,227	$1,227	$1,227	$14,724
	$150	$150	$150	$150	$150	$150	$150	$1,800

continued

CASH FLOW STATEMENT *(continued)* Year 3: September 1, 2009 to August 31, 2010					
	SEPT	**OCT**	**NOV**	**DEC**	**JAN**
Business License	$0	$0	$0	$0	$0
Computer	$0	$0	$0	$0	$0
Decor	$100	$100	$100	$100	$100
Electric	$175	$175	$175	$175	$175
Fax Machine	$0	$0	$0	$0	$0
Fixtures	$0	$0	$0	$0	$0
Gas and Water	$100	$300	$300	$300	$300
Insurance	$450	$450	$450	$450	$450
Inventory	$11,000	$12,000	$12,000	$22,000	$18,500
Kitchen Supplies	$0	$0	$0	$0	$0
Leasehold Imprvmnt.	$0	$0	$0	$0	$0
L&P Line of Credit	$0	$0	$0	$0	$0
Maintenance	$200	$200	$200	$200	$200
Mannequins	$0	$0	$0	$0	$0
Miscellaneous	$200	$200	$200	$200	$200
Office Furniture	$0	$0	$0	$0	$0
POS System Maint.	$100	$100	$100	$100	$100
Printer/Copier	$0	$0	$0	$0	$0
Rent	$1,747	$1,747	$1,747	$1,747	$1,747
Seating Area	$0	$0	$0	$0	$0
Signage	$0	$0	$0	$0	$0
Small Office Equipt.	$0	$0	$0	$0	$0
Software	$0	$0	$0	$0	$0
Staffing	$3,791	$3,791	$4,117	$4,117	$3,467
Staffing Taxes	$576	$576	$576	$576	$576
Stationery	$0	$0	$0	$0	$0
Steamer	$0	$0	$0	$0	$0
Supplies	$150	$150	$150	$150	$150
Telephone/Internet	$140	$140	$140	$140	$140
Travel	$1,460	$75	$75	$75	$75
Vacuum	$0	$0	$0	$0	$0
TOTAL FIXED EXPENSES	**$9,639**	**$8,454**	**$8,780**	**$8,780**	**$8,130**
TOTAL EXPENSES	**$25,271**	**$21,801**	**$22,637**	**$35,922**	**$27,977**
Cash on Hand (EOM)	$11,912	$13,421	$16,684	$11,842	$4,585
Operating Line of Credit	$0	$0	$0	$0	$0

	FEB	MARCH	APRIL	MAY	JUNE	JULY	AUGUST	TOTAL
	$0	$0	$0	$0	$0	$0	$0	$0
	$0	$0	$0	$0	$0	$0	$0	$0
	$100	$100	$100	$100	$100	$100	$100	$1,200
	$175	$175	$175	$175	$175	$175	$175	$2,100
	$0	$0	$0	$0	$0	$0	$0	$0
	$0	$0	$0	$0	$0	$0	$0	$0
	$300	$300	$100	$100	$100	$100	$100	$2,400
	$450	$450	$450	$450	$450	$450	$450	$5,400
	$10,500	$12,500	$10,500	$18,000	$15,500	$13,500	$10,500	$155,500
	$0	$0	$0	$0	$0	$0	$0	$0
	$0	$0	$0	$0	$0	$0	$0	$0
	$0	$0	$0	$0	$0	$0	$0	$0
	$200	$200	$200	$200	$200	$200	$200	$2,400
	$0	$0	$0	$0	$0	$0	$0	$0
	$200	$200	$200	$200	$200	$200	$200	$2,400
	$0	$0	$0	$0	$0	$0	$0	$0
	$100	$100	$100	$100	$100	$100	$100	$1,200
	$0	$0	$0	$0	$0	$0	$0	$0
	$1,747	$1,747	$1,747	$1,747	$1,747	$1,747	$1,747	$20,964
	$0	$0	$0	$0	$0	$0	$0	$0
	$0	$0	$0	$0	$0	$0	$0	$0
	$0	$0	$0	$0	$0	$0	$0	$0
	$0	$0	$0	$0	$0	$0	$0	$0
	$3,791	$3,467	$4,117	$4,117	$3,791	$3,467	$3,467	$45,500
	$576	$576	$576	$576	$576	$576	$576	$6,912
	$0	$0	$0	$0	$0	$0	$475	$475
	$0	$0	$0	$0	$0	$0	$0	$0
	$150	$150	$150	$150	$150	$150	$150	$1,800
	$140	$140	$140	$140	$140	$140	$140	$1,680
	$1,460	$75	$75	$75	$1,460	$75	$75	$5,055
	$0	$0	$0	$0	$0	$0	$0	$0
	$9,839	**$8,230**	**$8,780**	**$8,580**	**$9,639**	**$7,930**	**$7,905**	**$104,711**
	$22,196	**$24,687**	**$23,237**	**$27,927**	**$26,996**	**$22,777**	**$20,252**	**$290,680**
	$8,289	$12,092	$17,345	$17,908	$19,402	$19,935	$12,388	
	$0	$0	$0	$0	$0	$0	$0	

Cash Flow Statement Justifications
Year 1: September 1, 2007 to August 31, 2008

Cash on Hand	Samantha and Kate will invest a total of $22,000 into the business, an equal $11,000 supplied by each. This money will come from cash gifts, withdrawals from savings accounts, and from the selling off of personal stocks.
Decor	Each month, $100 has been allotted for fresh flowers, $25 of which will be used weekly.
Inventory	Inventory will be paid for by COD in the first year due to a lack of business credit, meaning that the inventory will be paid for the month it has been received. In the second and third years, once Bellissima has business credit, inventory will be paid for with billing/net 30 terms, thus meaning that inventory will be paid for the month after it has been received. Enough inventory will be bought to loosely keep the store full at a cost of $31,200. Inventory will dip some in the months of January, March, July, and August due to a slowdown in the number of social events. September and October inventory will be focused on work-oriented dressing.
Loan	Samantha and Kate will be requesting a loan of $40,000 to help cover start-up costs. For projection purposes, a 10.5 percent interest rate has been calculated for a period of 36 months. The loan will be used as follows: $31,200 Opening Inventory $4,500 Software $1,700 Signage $1,500 Computer $1,000 Décor $200 Small Office Equipment
Revolving Line of Credit	A line of credit for the amount of $30,000 has been requested. For projection purposes, interest accrues at a rate of 10.5 percent and will be paid monthly. With our calculations, we will not have to draw from this account, but the account remains ready for emergency purposes.

Pro Forma Income Statement
September 1, 2007 to August 31, 2008

Gross Sales	**$260,950**
Cost of Good Sold	
Beginning Inventory	$31,200
+ Purchases	$134,500
- Ending Inventory	$23,221
	$142,479

continued

Gross Profit$	**118,471**
Operating Expenses	
Staffing	$45,500
Staffing Taxes	$6,912
Rent	$20,964
Gas and Water	$2,400
Electricity	$2,100
Maintenance and Repairs	$2,400
Alarm Services	$1,200
Telephone and Internet	$1,680
Point of Sale System Maintenance/Updates	$1,200
Insurance	$5,400
Postage and Supplies	$1,800
Marketing and Advertising	$7,665
Travel and Entertainment	$4,830
Accounting and Bookkeeping	$3,900
Banking Services	$1,800
Depreciation and Amortization	$2,709
Miscellaneous	$2,400
Total	**$114,860**
Operating Profit	**$3,611**

Income Statement Justification
Year 1: Ending August 31, 2008

Gross Sales
We plan on having a conservative first year with our sales plan. Since we are a new store with a limited budget, we cannot afford to have a very aggressive marketing plan, which will in turn affect our gross sales. We will use our money wisely to increase sales but also rely on positive word of mouth. Our predicted gross sales for year one are $260,950.

Returns and Allowances
Since we are a new store, we will allow for returns in order to build positive relations with our clientele. If it becomes obvious that a customer is abusing our initial lenient policy (makes over two returns), we will not allow the customer to continue to return items. For customers who consistently make large purchases, we will give them a small percentage discount off the purchase to show our gratitude.

continued

Cost of Goods Sold
Since we will just be establishing our relationships with our vendors, we do not expect any specials or discounts. Before Bellissima builds up its Dun & Bradstreet Credit Rating, most vendors will give us payment terms of cash on delivery, or give us the option to place charges on a business credit card. We plan on purchasing $134,500 worth of merchandise for the first year.

Operating Expenses
Our operating expenses consist of everything we need to keep our store up and running. Included in the operating expenses:

- Accounting: We are hiring an accountant to help us make sure all of our numbers are correct, for payroll services, and to ensure we remain organized. This is especially important because the number one reason small businesses go under is poor record keeping.
- Advertising/Promotion: Our advertising budget is small for our first year of business because we do not have unlimited funding but we are using our dollars wisely. We developed our own Web site, have a pre-opening advertisement in *Richmond Magazine*, and plan on running advertisements in *Style Weekly*.
- Alarm Services: An alarm will protect our store 24 hours a day.
- Banking Services: We will need the bank to assist us with money orders and cashier checks. Also included in this number is the cost to process credit card transactions.
- Electric: We will be aware of our utility usage and try to keep costs down as much as possible.
- Gas and Water: We will be aware of our utility usage and try to keep costs down as much as possible.
- Insurance: We have liability insurance with the Erie Company. This is to ensure our financial safety in case something happens.
- Miscellaneous: This expense item is for anything that should happen to come up that we have not expected; we will have the miscellaneous expense for funding.
- POS System Maintenance: The POS system is used to keep track of all of our records for inventory and customers. It is vital that it is always working properly.
- Rent: Rent is nonnegotiable.
- Staffing: Both Kate and Samantha, the owner/managers, will be working full-time; one additional part-time employee will be hired.
- Staffing Taxes: Paid out at a rate of 14 percent.
- Supplies: This includes items such as business cards, stationery, envelopes, pricing tags, postcards, stapler, tape, paper, and so on.
- Telephone/Internet: We need a local line for customers to be able to contact us and a fax to place orders. The Internet is a staple item that any company needs in order to remain current with trends, communicate with vendors, and monitor our Web site.
- Travel: Three times a year we plan on traveling to trade shows to purchase new merchandise, meet new vendors, talk to other small boutique owners, and stay current with the trends.

Net Profit/Loss
After we compute the math, our store will have a profit of $3,611. Even though it is a small amount, the business will still be a success. We did not want to exaggerate the amount at which we think we could sell, yet would rather set a lower number and be realistic than set an absurdly high number and not be able to reach that goal. In following years our plan slowly becomes more aggressive as we build relationships with customers and vendors and have a larger budget.

Opening Pro Forma Balance Sheet
September 1, 2007

Current Assets		Current Liabilities	
Cash	$9,158.00	Current Portion of Long-Term Debt	$14,724.00
Inventory	$31,200.00	Line of Credit	$-
Prepaid Rent-Security Deposit	$1,747.00	**Total**	**$14,724.00**
Prepaid Utilities/Telephone/Alarm	$115.00		
Prepaid Marketing Expenses	$4,090.00	**Long-Term Liabilities**	
Supplies	$1,115.00	Long-Term Debt	$25,276.00
Total	**$47,425.00**	**Total**	**$40,000.00**
Other Assets		**Owners' Equity**	
Business Licenses	$230.00	Owners' Investment	$22,000.00
Leasehold Improvements Renovations	$800.00	Retained Earnings	$-
Furniture and Fixtures	$13,545.00	Plus: Net Income (Profit)	$-
		Less: Owners' Draw	$-
Total	**$14,575.00**	**Total**	**$22,000.00**
Total Assets	**$62,000.00**	**Total Liabilities and Owner Equity**	**$62,000.00**
Current Ratio	3.22		
Net Working Capital	$32,701.00		
Debt to Worth Ratio	1.82		

Closing Pro Forma Balance Sheet
August 31, 2008

Current Assets		Current Liabilities	
Cash	$7,058	Current Portion of Long-Term Debt	$14,724
Inventory	$23,221	Line of Credit	$0
Prepaid Rent-Security Deposit	$1,747	**Total**	$14,724
Prepaid Electricity	$115		
Prepaid Marketing Expenses	$4,090	Long-Term Liabilities	
Supplies	$1,590	Long-Term Debt	$10,552
Total	**$37,821**	**Total**	**$25,276**
Other Assets		**Owners' Equity**	
Business Licenses	$230	Owners' Investment	$22,000
Leasehold Improvements Renovations	$800	Retained Earnings	$0
Furniture and Fixtures	$14,745	Plus: Net Income (Profit)	$3,611
		Less: Owners' Draw	$0
Less: Depreciation and Amortization	$2,709	**Total**	**$25,611**
Total	**$13,066**		
Total Assets	**$50,887**	**Total Liabilities and Owner Equity**	**$50,887**
Current Ratio	$3		$0.30
Net Working Capital	$23,097		
Debt to Worth Ratio	$1		

Balance Sheet Justification
Year 1: Ending December 31, 2007

Assets

Our current assets consist of everything that we can turn into cash within one year. This includes our cash on hand at the end of December, which is $7,058, and our inventory in stock, which is worth $23,221.

Our fixed assets consist of the items purchased for long-term use, or use over one year. This includes fixtures, furniture, and computer valued at $14,745. We plan on replacing these items every five years.

Liabilities

To pay the $40,000 loan in full in 36 months, including interest, consists of $44,200. Current liabilities consist of the portion of the loan that we will have to repay in the first year, which is $14,724. The remainder of our long-term debt consists of the portion of the entire loan that we will have to repay in the following years. We did not take any draws from the line of credit and therefore have no payments toward it.

Total Equity and Net Worth

After determining the total amount of assets we possess and subtracting the total amount of liabilities we have accrued, we calculated that at the end of its first calendar year Bellissima will be worth $25,611.

Key Ratios
First Year of Operation

	Income Statement	Opening Balance Sheet	Closing Balance Sheet	
Gross Profit Ratio				
Gross Profit	$118,471.30			
Sales	$260,950.00			
	0.45			
Net Profit Ratio				
Net Profit	$3,611.30			
Sales	$260,950.00			
	0.01			
Current Ratio		**3.22**	**2.57**	
Net Working Capital		**$32,701.00**	**$23,097.00**	
Debt to Worth Ratio		**1.82**	**0.99**	
Return on Equity				
Owners Equity		$22,000.00	$25,611.30	
Average Owners Equity				$23,805.65
Net Income				$3,611.30
Return on Debt				
Debt		$40,000.00	$25,276.00	
Average Debt				$32,638.00
Net Income				$3,611.30
Return on Assets				
Assets		$62,000.00	$50,887.00	
Average Assets				$56,443.50
Net Income				$3,611.30
				0.06

CORE VENDORS			
DESIGNER	**ADDRESS**	**CONTACT NUMBER**	**PRICE POINT**
DRESSES			
A.B.S. by Allen Schwartz	525 Seventh Avenue, 14th floor, NYC 10018	T: 212.398.0330 F: 212.840.0092	$250, $300, $350
AdamPlusEve	450 West 15th Street, 4th floor, NYC 10011	T: 212.675.2022 F: 212.675.2893	$275, $300, $350
Chaiken	580 Broadway, Suite 400, NYC 10012	T: 212.334.3501 F: 212.334.3504	$200, $250
CK Bradley	146 East 74th Street, NYC 10021	T: 212.988.7999 F: 212.3962988	$175, $250, $300
Cynthia Steffe	550 Seventh Avenue, 10th floor, NYC 10018	T: 212.403.6200 F: 212.302.1254	$200–$250
Diane von Furstenburg	389 West 12th Street, NYC 10014	T: 212.741.6607 F: 212.741.8273	$200, $275, $350
Fashionista	231 West 39th Street, Suite 801, NYC 10018	T: 212.354.0095 F: 212.354.0080	$125, $175
Shoshanna	231 West 39th Street, Suite 422, NYC 10018	T: 212.719.3601 F: 212.719.0745	$300, $350
Tailor New York	234 West 39th Street, 8th floor, NYC 10018	T: 212.840.1871 F: 212.840.1877	$125, $175
Tory Burch	99 Madison Avenue, 12th floor, NYC 10016	T: 646.723.6622 F: 646.514.4135	$250, $300, $350
Trina Turk	3025 West Mission Road, Alhambra CA 91803	T: 626.458.7768 F: 626.457.8439	$250, $300, $350
HANDBAGS			
Hype	910 South Los Angeles Street, LA, CA 90015	T: 213.489.9605 F: 213.489.2639	$75, $100
Jazzd	P.O. Box 153697, San Diego, CA 92195	T: 619.469.4778 F: 619. 741.4892	$25, $35, $50
Lulu Guinness	934 Bleecker Street, NY, 10014	T: 212.367.2120 F: 212.367.2140	$100, $150
JEWELRY			
A.V. Max	16 West 36th Street, NYC 10018	T: 212.502.0960	$35, $50
Ben-Amun	246 West 38th Street, Suite 12A, NYC 10018	T: 212.944.6480 F: 212.944.9625	$75, $125
Gerard Yosca	39 West 38th Street, 10th Floor, NYC 10018	T: 212.302.4349 F: 212.768.8836	$75, $125
Local Vendors	Variable	Variable	$25, $35, $50

OPENING INVENTORY						
CATEGORY	**PRICE POINT, COST**	**PRICE POINT, RETAIL**	**PERCENT AT PRICE POINT**	**PIECES PER PRICE POINT**	**TOTAL AT COST**	**TOTAL AT RETAIL**
DRESSES	$56	$125	14%	78	$4,366	$9,614
	$79	$175	16%	63	$4,990	$10,987
	$112	$250	16%	45	$4,990	$10,987
	$135	$300	14%	32	$4,366	$9,614
	$158	$350	10%	20	$3,119	$6,867
TOTAL			**70%**	**238**	**$21,831**	**$48,070**
JEWELRY	$11	$25	5%	142	$1,559	$3,434
	$16	$35	4%	78	$1,247	$2,747
	$23	$50	2%	27	$624	$1,373
	$34	$75	2%	18	$624	$1,373
	$56	$125	2%	11	$624	$1,373
TOTAL			**15%**	**276**	**$4,678**	**$10,301**
HANDBAGS	$23	$50	3%	41	$936	$2,060
	$34	$75	5%	46	$1,559	$3,434
	$45	$100	3%	21	$936	$2,060
	$68	$150	2%	9	$624	$1,373
	$90	$200	2%	7	$624	$1,373
TOTAL			**15%**	**123**	**$4,678**	**$10,301**
STORE TOTAL			**100%**	**638**	**$31,187**	**$68,671**

LOCATION FORMS AND FLOOR PLAN

Store Location Fact List

Address	5710 Patterson Avenue, Richmond, Virginia 23226 On the northeast corner of Patterson and Libbie Avenues
Square Footage	Total: 1397.5 square feet (width: 25.8 feet, length: 54.15 feet) Restroom: 26 square feet (width: 5.0 feet, length: 5.2 feet) Storage closet: 26 square feet (width: 5.0 feet, length: 5.2 feet) Storage/office room: 336.9 square feet (width: 15.11 feet, length: 22.3 feet) Selling space: 800 square feet (space does not include storage closet, storage/office room, bathroom, or rear hallway) Square footage as provided by the Jones Realty Group, excluding calculation of selling space
Location of Entrances/Exits	The front entrance consists of a single, half-window door located in the front right corner of the store, facing Patterson Avenue. The rear entrance consists of a single, half-window door located in the rear, left-hand corner of the store. It faces the alley and a small private parking lot.
Location of Windows	The sole window in the store is a plate-glass window in the front, composing the majority of the wall space not occupied by the front entrance.
Monthly Rent	$15/sq.ft yearly, or $1,747 per month.
Terms of Lease	Flat rate, with no cost increase in second year. No subletting. The lease would be a triple net lease, meaning that the lessee is fully responsible for all costs associated with the property being leased.
Common Area Fees	No.
Security Deposit	One month's rent, or $1,747.
Realtor Fees	No, we contacted the leaser, Jones Realty Group.
Cost of Utilities	Electricity, power, and water to be paid for by lessee, per triple net lease standards.
Renovations Allowed	Interior: consisting of nonpermanent structures. Exterior: paint, awning. Jones Realty Group is not willing to pay for any improvements or renovations.

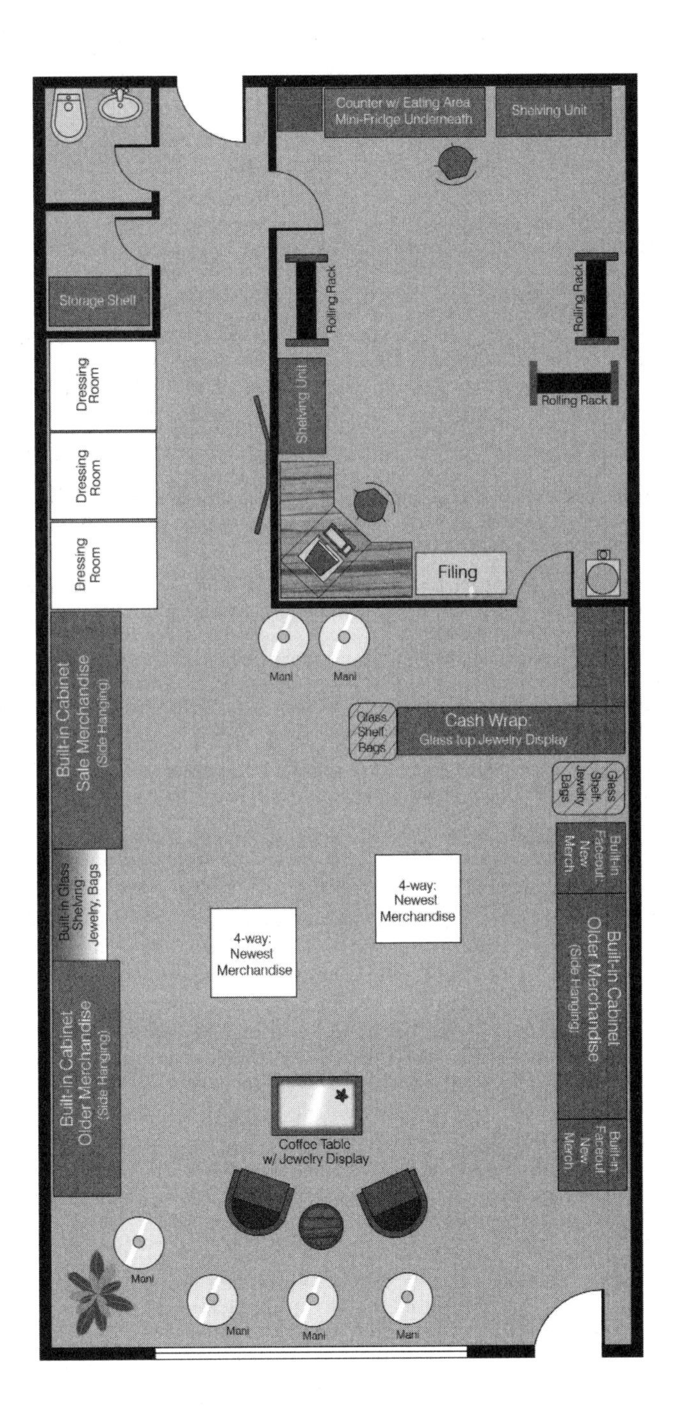

Counter w/ Eating Area
Mini-Fridge Underneath
Shelving Unit
Storage Shelf
Rolling Rack
Rolling Rack
Rolling Rack
Shelving Unit
Dressing Room
Dressing Room
Dressing Room
Filing
Mani
Mani
Built-in Cabinet
Sale Merchandise
(Side Hanging)
Glass Shelf Bags
Cash Wrap:
Glass top Jewelry Display
Glass Shelf Jewelry Bags
Built-in Faceout: New Merch.
4-way:
Newest
Merchandise
4-way:
Newest
Merchandise
Built-in Glass Shelving: Jewelry, Bags
Built-in Cabinet
Older Merchandise
(Side Hanging)
Built-in Cabinet
Older Merchandise
(Side Hanging)
Coffee Table
w/ Jewelry Display
Built-in Faceout New Merch.
Mani
Mani
Mani
Mani

Commercial Lease

This lease is made between Jones Realty Construction Corporation, herein called Lessor, and Kate Aliberti, and Samantha Baum, herein called Lessee.

Lessee hereby offers to lease from Lessor the premises situated in the County of Henrico, State of Virginia, described as Retail Location, 5710 Patterson Avenue upon the following TERMS and CONDITIONS.

1. Term and Rent. Lessor demises the above premises for a term of 2 years, commencing on September 1, 2007 and terminating on July 31, 2009, or sooner as provided herein at the annual rental of $15 per square foot, per year, with the square footage of the above location being 1397.5 feet squared, therefore resulting in a rent per month of one thousand seven hundred and forty-seven dollars ($1,747) payable in equal installments in advance on the first day of each month for that month's rental, during the term of this lease. All rental payments shall be made to Lessor, at the address specified above.
2. Use. Lessee shall use and occupy the premise for the purpose of a retail clothing store. The premises shall be used for no other purpose. Lessor represents that the premises may lawfully be used for such purpose.
3. Care and Maintenance of Premises. Lessee acknowledges that the premises are in good order and repair, unless otherwise indicated herein. Lessee shall, at his or her own expense and at all times, maintain the premises in good and safe condition, including plate glass, electrical wiring, plumbing and heating installations and any other system or equipment upon the premises, and shall surrender the same at termination hereof, in as good condition as received, normal wear and tear excepted. Lessee shall be responsible for all repairs required, excepting the roof, exterior walls, and structural foundations.
4. Alterations. Lessee shall not, without first obtaining the written consent of Lessor, make any alterations, additions, or improvements, in, to, or about the premises.
5. Ordinances and Statutes. Lessee shall comply with all statutes, ordinances, and requirements of all municipal, state, and federal authorities now in force, or which may hereafter be in force, pertaining to the premises, occasioned by or affecting the use thereof by Lessee.
6. Assignment and Subletting. Lessee shall not assign this lease or sublet any portion of the premises without prior written consent of the Lessor, which shall not be unreasonably withheld. Any such assignment or subletting without consent shall be void and, at the option of the Lessor, may terminate this lease.
7. Utilities. All applications and connections for necessary utility services on the demised premises shall be made in the name of Lessee only, and Lessee shall be solely liable for utility charges as they become due, including those for sewer, water, gas, electricity, and telephone services.
8. Entry and Inspection. Lessee shall permit Lessor or Lessor's agents to enter upon the premises at reasonable times and upon reasonable notice, for the purposes of inspecting the same, and will permit Lessor at any time within sixty (60) days prior to the expiration of this lease, to place upon the premises any usual "To Let" or "For Lease" signs, and permit persons desiring to lease the same to inspect the premises thereafter.
9. Possession. If Lessor is unable to deliver possession of the premises at the commencement hereof, Lessor shall not be liable for any rent until possession is delivered. Lessee may terminate this lease if possession is not delivered within days of the commencement of the term hereof.

10. Indemnification of Lessor. Lessor shall not be liable for any damage or injury to Lessee, or any other person, or to any property, occurring on the demised premises or any part thereof, and Lessee agrees to hold Lessor harmless from any claim for damages, no matter how caused.

11. Insurance. Lessee, at his expense, shall maintain plate glass and public liability insurance including bodily injury and property damage insuring Lessee and Lessor with minimum coverage as follows:

    Lessee shall provide Lessor with a Certificate of Insurance showing Lessor as additional insured. The Certificate shall provide for a ten-day written notice to Lessor in the event of cancellation or material change of coverage. To the maximum extent permitted by insurance policies which may be owned by Lessor or Lessee, Lessee and Lessor, for the benefit of each other, waive any and all rights of subrogation which might otherwise exist.

12. Eminent Domain. If the premises or any part thereof or any estate therein, or any other part of the building materially affecting Lessee's use of the premise, shall be taken by eminent domain, this lease shall terminate on the date when title vests pursuant to such taking. The rent, and any additional rent, shall be apportioned as of the termination date, and any rent paid for and period beyond that date shall be repaid to Lessee. Lessee shall not be entitled to any part of the award for such taking or any payment in lieu thereof, but Lessee may file a claim for any taking of fixtures and improvements owned by Lessee, and for moving expenses.

13. Destruction of Premises. In the event of a partial destruction of the premises during the term hereof, from any cause, Lessor shall forthwith repair the same, provided that such repairs can be made within sixty (60) days under existing governmental laws and regulations, but such partial destruction shall not terminate this lease, except that Lessee shall be entitled to a proportionate reduction of rent while such repairs are being made, based upon the extent to which making the repairs cannot be made within sixty (60) days, Lessor, at his or her option, may make the same within a reasonable time, this lease continuing in effect with the rent proportionately abated as aforesaid, and in the event that Lessor shall not elect to make such repairs which cannot be made within sixty (60) days, this lease may be terminated at the option of either party. In the event that the building in which the demised premises may be situated is destroyed to an extent of not less than one-third of the replacement cost, Lessor may elect to terminate this lease whether the demised premises be injured or not. A total destruction of the building in which the premises may be situated shall terminate this lease.

14. Lessor's Remedies on Default. If Lessee defaults in the payment of rent, or any additional rent, or defaults in the performance of any of the other covenants or conditions hereof, Lessor may give Lessee notice of such default and if Lessee does not cure any such default within ten (10) days, after the giving of such notice (or if such other default is of such nature that it cannot be completely cured within such period, if Lessee does not commence such curing within such ten (10) days and thereafter proceed with reasonable diligence and in good faith to cure such default, then Lessor may terminate this lease on not less than fourteen (14) days' notice to Lessee. On the date specified in such notice the term of this lease shall terminate, and Lessee shall then quit and surrender the premises to Lessor, but Lessee shall remain liable as hereinafter provided. If this lease shall have been so terminated by Lessor, Lessor may at any time thereafter resume possession of the premises by any lawful means and remove Lessee or other occupants and their effects. No failure to enforce any term shall be deemed a waiver.

15. Security Deposit. Lessee shall deposit with Lessor on the signing of this lease the sum of one month's rent, or one thousand seven hundred and forty-seven dollars ($1,747) as

continued

security deposit for the performance of Lessee's obligations under this lease, including without limitation the surrender of possession of the premises to Lessor as herein provided. If Lessor applies any part of the deposit to cure any default of Lessee, Lessee shall on demand deposit with Lessor the amount so applied so that Lessor shall have the full deposit on hand at all times during the term of this lease.

16. Tax Increase. In the event there is any increase during any year of the term of this lease in the City, County or State real estate taxes over and above the amount of such taxes assessed for the tax year during which the term of this lease commences, whether because of increased rate or valuation, Lessee shall pay to Lessor upon presentation of paid tax bills an amount equal to seventy-five percent (75%) of the increase in taxes upon the land and building in which the leased premises are situated. In the event that such taxes are assessed for a tax year extending beyond the term of the lease, the obligation of Lessee shall be proportionate to the portion of the lease term included in such year.

17. Common-Area Expenses. In the event the demised premises are situated in a shopping center or in a commercial building in which there are common areas, Lessee agrees to pay his pro rata share of maintenance, taxes, and insurance for the common area.

18. Attorney's Fees. In case suit should be brought for recovery of the premises, or for any sum due hereunder, or because of any act which may arise out of the possession of the premises, by either party, the prevailing party shall be entitled to all costs incurred in connection with such action, including reasonable attorney's fee.

19. Notices. Any notice which either party may, or is required to give, shall be given mailing same, postage prepaid, to Lessee at the premises, or Lessor at the address shown below, or at such other places as may be designated by the parties from time to time.

20. Heirs, Assigns, Successors. This lease is binding upon and inures to the benefit of the heirs, successors in interest to the parties.

21. Option to Renew. Provided that Lessee is not in default in the performance of this lease, Lessee shall have the option to renew the lease for an additional term of twelve (12) months commencing at the expiration of the initial lease term. All of the terms and conditions of the lease shall apply during the renewal term. The option shall be exercised by written notice given to Lessor not less than thirty (30) days prior to the expiration of the initial lease term. If notice is not given in the manner provided herein within the time specified, this option shall expire.

22. Subordination. This lease is and shall be subordinated to all existing and future liens and encumbrances against the property.

23. Entire Agreement. The foregoing constitutes the entire agreement between the parties and may be modified only in a writing signed by both parties. The following Exhibits, if any, have been made a part of this lease before the parties' execution hereof:

Signed this day of ______________ July 1, 2007

Lessor: ______________ John Jones ______________

Lessee 1: ______________ Kate Aliberti ______________

Lessee 2: ______________ Samantha Baum ______________

MONTHLY MARKETING PLAN Year 1: September 2007 to August 2008							
	CUE	**STYLE WEEKLY**	**RICHMOND MAGAZINE**	**RICHMOND BRIDE**	**WEB SITE**	**MAILER**	**TOTAL**
PRE-OPENING		$510	$2,100		$20		**$2,630**
SEPTEMBER	$510			$2,100	$20		**$2,630**
OCTOBER					$20		**$20**
NOVEMBER					$20		**$20**
DECEMBER	$510	$510			$20	$675	**$1,715**
JANUARY					$20		**$20**
FEBRUARY					$20		**$20**
MARCH	$510			$2,100	$20		**$2,630**
APRIL					$20		**$20**
MAY					$20		**$20**
JUNE	$510				$20		**$530**
JULY					$20		**$20**
AUGUST					$20		**$20**
TOTAL	**$2,040**	**$510**	**$0**	**$4,200**	**$240**	**$675**	**$7,665**

PROFILE FOR ZIP CODE 23226		
POPULATION		
Total population	18,021	Out of the 9,570 women that live in the area, we estimate that approximately 4,600 will be in our target market, aged 18 to 45. The average age in the zip code is 33.5, which will fall into our target age group. Of the 11,503 people aged 25 and older, 6,108 are women. Since 52.1 percent will have obtained a bachelor's degree or higher; this further breaks our segmentation down into 3,182 women that are aged 25 and
15 to 19	1,833	
20 to 24	2,317	
25 to 34	2,852	
35 to 44	2,270	
45 to 54	2,031	
55 to 59	784	
Median age	33.5	
Female	9,570	
Female over 18	8,166	

continued

PROFILE FOR ZIP CODE 23226 *(continued)*	
HOUSEHOLDS	
Total housing units	7,333
Total households	7,004
Average household size	2.14
Female over 18	2.79
EDUCATIONAL ATTAINMENT	
Population 25 and older	11,503
High school graduate	2,333
Some college, no degree	1,333
Associate degree	427
Bachelor's degree	3,333
Graduate or professional degree	2,333
High school graduate or higher, in percent	90.5
Bachelor's degree or higher, in percent	52.1
INCOME	
HOUSEHOLDS	7,013
Less than $10,000	342
$10,000 to $14,999	223
$15,000 to $24,999	755
$25,000 to $34,999	973
$35,000 to $49,999	1,173
$50,000 to $74,999	1,349
$75,000 to $99,999	677
$100,000 to $149,999	743
$150,000 to $199,999	223
$200,000 or more	555
Median household income	$50,519
FAMILIES	3,920
Less than $10,000	59
$10,000 to $14,999	50
$15,000 to $24,999	266
$25,000 to $34,999	317
$35,000 to $49,999	619
$50,000 to $74,999	915
$75,000 to $99,999	424
$100,000 to $149,999	594
$150,000 to $199,999	188
$200,000 or more	488
Median household income	$66,505

older that have completed college. We feel that educational attainment is important because it leads to a higher income later in life.

Following the assumption that college-educated people make more money, it is safe to assume that those people with a household or family income of $50,000 or more have attended college. Out of the 7,013 total households in the zip code, 3,547 make over $50,000. Since the average household includes approximately two people, there are 1,773 women. Out of the 3,920 families in the zip code, 2,609 make over $50,000. Since the average household contains two adults and one child, 869 are women. Adding the two numbers together leaves 2,642 women.

In the zip code 23226, there are 2,642 women in our target group.

Data Source: U.S. Census Bureau

PROFILE FOR ZIP CODE 23221

POPULATION	
Total population	13,635
15 to 19	376
20 to 24	1,520
25 to 34	3,621
35 to 44	2,154
45 to 54	1,824
55 to 59	598
Median age	34.6
Female	7,082
Female over 18	6,303
HOUSEHOLDS	
Total housing units	7,841
Total households	7,496
Average household size	1.80
Female over 18	2.65
EDUCATIONAL ATTAINMENT	
Population 25 and older	10,267
High school graduate	1,251
Some college, no degree	1,903
Associate degree	321
Bachelor's degree	3,988
Graduate or professional degree	2,095
High school graduate or higher, in percent	93.1
Bachelor's degree or higher, in percent	59.2
INCOME	
HOUSEHOLDS	7,401
Less than $10,000	601
$10,000 to $14,999	422
$15,000 to $24,999	977
$25,000 to $34,999	1,147
$35,000 to $49,999	1,195
$50,000 to $74,999	1,278
$75,000 to $99,999	602
$100,000 to $149,999	634
$150,000 to $199,999	224
$200,000 or more	321
Median household income	$39,943

Out of the 7,082 women that live in the area, we estimate that approximately 3,786 will be in our target market, aged 18 to 45. The average age in the zip code is 34.6, which will fall into our target age group.

Of the 10,267 people aged 25 and older, 5,328 are women. Since 59.2 percent will have obtained a bachelor's degree or higher, this further breaks our segmentation down into 3,154 women that are aged 25 and older that have completed college. We feel that educational attainment is important because it leads to a higher income later in life.

Following the assumption that college-educated people make more money, it is safe to assume that those people with a household or family income of $50,000 or more have attended college. Out of the 7,401 total households in the zip code, 3,059 make over $50,000. Since the average household includes approximately two people, there are 1,529 women. Out of the 2,567 families in the zip code, 1,675 make over $50,000. Since the average household contains two adults and one child, 558 are women. Adding the two numbers together leaves 2,087 women.

In the zip code 23221, there are 2,087 women in our target group.

Data Source: U.S. Census Bureau

continued

PROFILE FOR ZIP CODE 23221 *(continued)*		
INCOME ***(continued)***		
FAMILIES	2,567	
Less than $10,000	66	
$10,000 to $14,999	58	
$15,000 to $24,999	154	
$25,000 to $34,999	229	
$35,000 to $49,999	375	
$50,000 to $74,999	508	
$75,000 to $99,999	348	
$100,000 to $149,999	416	
$150,000 to $199,999	138	
$200,000 or more	265	
Median household income	$69,358	

<table>
<tr><th colspan="3">PROFILE FOR ZIP CODE 23229</th></tr>
<tr><th>POPULATION</th><th></th><th></th></tr>
<tr><td>Total population</td><td>31,051</td><td rowspan="21">Out of the 16,344 women that live in the area, we estimate that approximately 10,012 will be in our target market, aged 18 to 45. The average age in the zip code is 40.6, which will fall into our target age group.

Of the 21,800 people aged 25 and older, 11,466 are women. Since 50.5 percent will have obtained a bachelor's degree or higher, this further breaks our segmentation down into 5,790 women that are aged 25 and older that have completed college. We feel that educational attainment is important because it leads to a higher income later in life.

Following the assumption that college-educated people make more money, it is safe to assume that those people with a household or family income of $50,000 or more have attended college. Out of the 12,866 total households in the zip code, 7,410 make over $50,000. Since the average household includes approximately two people, there are 3,705 women. Out of the 8,734 families in the zip code, 5,977 make over $50,000</td></tr>
<tr><td>15 to 19</td><td>1,787</td></tr>
<tr><td>20 to 24</td><td>1,519</td></tr>
<tr><td>25 to 34</td><td>3,698</td></tr>
<tr><td>35 to 44</td><td>4,795</td></tr>
<tr><td>45 to 54</td><td>4,722</td></tr>
<tr><td>55 to 59</td><td>1,629</td></tr>
<tr><td>Median age</td><td>40.6</td></tr>
<tr><td>Female</td><td>16,344</td></tr>
<tr><td>Female over 18</td><td>12,758</td></tr>
<tr><th>HOUSEHOLDS</th><th></th></tr>
<tr><td>Total housing units</td><td>13,213</td></tr>
<tr><td>Total households</td><td>12,877</td></tr>
<tr><td>Average household size</td><td>2.40</td></tr>
<tr><td>Female over 18</td><td>2.93</td></tr>
<tr><th>EDUCATIONAL ATTAINMENT</th><th></th></tr>
<tr><td>Population 25 and older</td><td>21,800</td></tr>
<tr><td>High school graduate</td><td>3,652</td></tr>
<tr><td>Some college, no degree</td><td>4,538</td></tr>
<tr><td>Associate degree</td><td>968</td></tr>
<tr><td>Bachelor's degree</td><td>6,697</td></tr>
</table>

continued

PROFILE FOR ZIP CODE 23229 *(continued)*	
EDUCATIONAL ATTAINMENT *(continued)*	
Graduate or professional degree	4,307
High school graduate or higher, in percent	92.5
Bachelor's degree or higher, in percent	50.5
INCOME	
HOUSEHOLDS	12,866
Less than $10,000	449
$10,000 to $14,999	413
$15,000 to $24,999	1,091
$25,000 to $34,999	1,414
$35,000 to $49,999	2,109
$50,000 to $74,999	2,786
$75,000 to $99,999	1,785
$100,000 to $149,999	1,373
$150,000 to $199,999	627
$200,000 or more	839
Median household income	$57,251
FAMILIES	8,734
Less than $10,000	140
$10,000 to $14,999	131
$15,000 to $24,999	414
$25,000 to $34,999	757
$35,000 to $49,999	1,315
$50,000 to $74,999	1,881
$75,000 to $99,999	1,490
$100,000 to $149,999	1,207
$150,000 to $199,999	588
$200,000 or more	811
Median household income	$69,944

or more have attended college. Out of the 12,866 total households in the zip code, 7,410 make over $50,000. Since the average household includes approximately two people, there are 3,705 women. Out of the 8,734 families in the zip code, 5,977 make over $50,000. Since the average household contains two adults and one child, 1,992 are women. Adding the two numbers together leaves 5,697 women.

In the zip code 23229, we estimate that there are 5,697 women in our target group.

Data Source: U.S. Census Bureau

PROFILE FOR ZIP CODE 23220	
POPULATION	
Total population	29,864
15 to 19	3,270
20 to 24	5,438
25 to 34	6,038
35 to 44	3,760
45 to 54	3,173
55 to 59	1,071
Median age	28.9
Female	15,333

Out of the 15,333 women that live in the area, we estimate that approximately 15,236 will be in our target market, aged 18 to 45. The average age in the zip code is 28.9, which will fall into our target age group.

Of the 17,888 people aged 25 and older, 9,176 are women. Since 39.5 percent will have obtained a bachelor's degree or higher, this further breaks

continued

PROFILE FOR ZIP CODE 23220 *(continued)*		
POPULATION *(continued)*		
Female over 18	13,366	our segmentation down into 3,624 women that are aged 25 and older that have completed college. We feel that educational attainment is important because it leads to a higher income later in life. Following the assumption that college-educated people make more money, it is safe to assume that those people with a household or family income of $50,000 or more have attended college. Out of the 13,416 total households in the zip code, 3,414 make over $50,000. Since the average household includes approximately two people, there are 1,707 women. Out of the 4,736 families in the zip code, 1,104 make over $50,000. Since the average household contains two adults and one child, 368 are women. Adding the two numbers together leaves 2,075 women. In the zip code 23220, we estimate that there are 2,075 women in our target group. *Data Source: U.S. Census Bureau*
HOUSEHOLDS		
Total housing units	14,790	
Total households	13,396	
Average household size	1.95	
Female over 18	2.81	
EDUCATIONAL ATTAINMENT		
Population 25 and older	17,888	
High school graduate	3,095	
Some college, no degree	3,226	
Associate degree	500	
Bachelor's degree	4,415	
Graduate or professional degree	2,646	
High school graduate or higher, in percent	77.6	
Bachelor's degree or higher, in percent	39.5	
INCOME		
HOUSEHOLDS	13,416	
Less than $10,000	3,000	
$10,000 to $14,999	1,451	
$15,000 to $24,999	2,024	
$25,000 to $34,999	1,677	
$35,000 to $49,999	1,861	
$50,000 to $74,999	1,543	
$75,000 to $99,999	696	
$100,000 to $149,999	633	
$150,000 to $199,999	243	
$200,000 or more	299	
Median household income	$26,557	
FAMILIES	4,736	
Less than $10,000	736	
$10,000 to $14,999	312	
$15,000 to $24,999	647	
$25,000 to $34,999	544	
$35,000 to $49,999	705	
$50,000 to $74,999	688	
$75,000 to $99,999	358	
$100,000 to $149,999	395	
$150,000 to $199,999	157	
$200,000 or more	194	
Median household income	$36,792	

PERSONNEL AND MANAGEMENT PLAN DOCUMENTS

PAYROLL BUDGET FOR WEEK OF: November 11–17, 2007															
SALES ASSOCIATE	SUNDAY		MONDAY		TUESDAY		WEDNESDAY		THURSDAY		FRIDAY		SATURDAY		TOTAL HOURS
	TIME	HOURS	TIME	HOURS	TIME	HOURS	TIME	HOURS	TIME	HOURS	TIME	HOURS	TIME	HOURS	
Samantha			10-6	8	10-6	8	10-6	8	10-6	8	10-6	8			40
Kate					10-6	8	10-6	8	10-6	8	10-6	8	10-6	8	40
PT associate			11-6	7									10-6	8	15
TOTAL		**0**		**15**		**16**		**16**		**16**		**16**		**16**	**95**

Weekly sales budget:	$6,000 $5,436
Weekly staff hours budget:	95
Average hourly rate:	$10
Weekly payroll budget:	$950
Weekly payroll to sales percent:	15.8% 12.8%
Hours open:	M-Sa 10-6

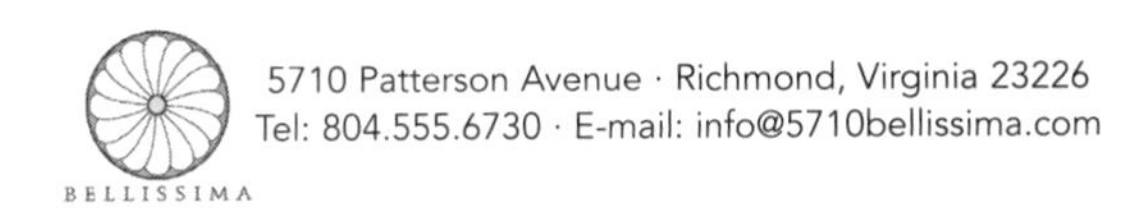

SALES ASSOCIATE POSITION

Overview:
Located on the corner of Libbie and Patterson Avenues, Bellissima offers a variety of exclusive dresses, handbags, and jewelry for trendy women in a comfortable setting. Exceptional customer service and a personal stylist will make finding the perfect dress for any occasion fun and easy, whether in-house or around town.

Responsibilities:
- Greet customers and ascertain what each customer wants or needs.
- Open and close cash registers, performing tasks such as counting money, separating charge slips, coupons, and vouchers, balancing cash drawers, and making deposits.
- Maintain knowledge of current sales and promotions, policies regarding payment and exchanges, and security practices.
- Compute sales prices, total purchases, and receive and process cash or credit payment.
- Maintain records related to sales.
- Watch for and recognize security risks and thefts, and know how to prevent or handle these situations.
- Recommend, select, and help locate or obtain merchandise based on customer needs and desires.
- Answer questions regarding the store and its merchandise.
- Describe merchandise and its care to customers.
- Aid in ticketing, arranging, and displaying merchandise to promote sales.
- Complete sales transactions on register.
- Demonstrate use or operation of merchandise.
- Clean shelves, counters, and tables.
- Exchange merchandise for customers and accept returns.
- Bag or package purchases, and wrap gifts.
- Help customers try on or fit merchandise.
- Participate in inventory assessment, held every six months.
- Prepare new merchandise for sales floor.
- Requirements:
- Ability to stand for up to five hours at a time
- Ability to compute basic math
- Ability to lift moderately heavy boxes

Business Resume

KATE ALIBERTI

125 MAIN STREET · RICHMOND VIRGINIA 23221
PHONE 804.555.3946 · EMAIL ONEILKA@VCU.EDU

WORK EXPERIENCE

Anthropologie 2005 to Present
Richmond, Virginia
Store Manager

- A national chain with a boutique feel offering a wide variety of women's clothing, jewelry, and accessories for the home.
- Utilize and manage the use of weekly sales reports to track, analyze, and communicate business results and determine strategies to maximize sales.
- Partner with the Visual Manager in planning, scheduling, and executing an innovative visual environment.
- Manage and train store employees.
- Manage all aspects of store operational controls.

A.R. Bevans 2001 to 2004 Richmond, Virginia
Manager and Sales Representative

- A small boutique specializing in contemporary women's clothing and accessories, catering to an upscale clientele that enjoys a personalized shopping experience.
- Managed the sales floor and learned the intricacies of running a retail business on a daily basis.
- Organized special functions for the store, resulting in capability to plan events from initial idea through post-function cleanup, including advertising promotions.
- Created enticing visual displays and signage, on a weekly basis, to draw customers into the store and encourage customer purchases.
- Offered personalized customer service, which led to a higher daily sales goal and increased customer loyalty.

EDUCATION

Virginia Commonwealth University May 2004 Graduate
Richmond, Virginia
B.A. Fashion Merchandising, Major
General Business, Minor

- Participated in semester-long simulations, developing a department store buying plan and internationally importing denim for a department store.
- Attended a line development class in Guatemala, resulting in a knowledge of design and construction of clothing. The line of fair-trade clothing developed is now sold in a local Richmond boutique.
- Forecasted fashion trends for one year in advance, with the presentation being posted on the department Web site.
- Learned features of proper product promotion through various marketing classes.

RELATED EXPERIENCE AND SKILLS

- Served as a visiting stylist for Richmond.com, highlighting spring trends from local boutiques.
- Computer skills in Microsoft Word, Excel, PowerPoint, and Outlook; Adobe Photoshop CS and Illustrator CS.
- Basic knowledge of French.

Personal Financial Statements

KATE ALIBERTI

ASSETS

Cash and Cash Equivalents			
Savings & Loan Bank	Checking Acct. #123987456		$1,500
Savings & Loan Bank	Savings Acct. #123654789		$3,700
		TOTAL	**$5,200**
Cash On Hand			
Cash at home			$500
		TOTAL	**$500**
Marketable Securities			
Altria Group, Inc. (MO)	50 shares on NYSE		$4,300
Target Corporation (TGT)	100 shares on NYSE		$6,200
		TOTAL	**$10,500**
Real Estate			
Personal Residence	125 Main Street, Richmond, VA 23221 1408 sq. ft.; 2 bedroom, 1.5 baths; single-family 2-floor brick row house, attached on one side; built 1914; assessed at $256,000, joint owned with husband		$128,000
		TOTAL	**$128,000**
Other Assets			
Cash gift from parents			$5,000
		TOTAL	**$5,000**
Personal Property			
1997 BMW Z3, excellent condition			$10,200
1890s Mahogany dining table			$1,100
Original Matisse lithograph, with authenticity papers			$600
Antique solid-core silver set			$7,000
14 place settings, 9 pieces per set, plus serving utensils			
Spode bone china			$4,000
8-place settings, 6 pieces per set			
Miscellaneous antique furniture			$3,000
Miscellaneous oriental rugs			$2,500
Other household items			$5,000
		TOTAL	**$33,400**
		ASSETS TOTAL	**$182,600**

KATE ALIBERTI

LIABILITIES AND NET WORTH

Credit Cards			
VISA (MBNA)			$1,000
		TOTAL	**$1,400**
Loans Secured By Real Estate			
CitiBank	30-year fixed-rate loan, maturing in 2035; $185,000 to be paid in monthly installments, split with husband		$92,500
		TOTAL	**$92,500**
		LIABILITIES TOTAL	**$93,500**
		ASSETS TOTAL	**$182,600**
		LIABILITIES TOTAL	**($93,500)**
		NET WORTH TOTAL	**$89,100**
		LIABILITIES AND NET WORTH TOTAL	**$182,600**

KATE ALIBERTI

ANNUAL INCOME			
Gross Salary and Wages			
Anthropologie	Store Manager		$43,000
		TOTAL	**$43,000**
Dividends and Interest			
Altria Group, Inc.	Dividends on 50 shares		$170
Target Corporation	Dividends on 100 shares		$50
SunTrust Bank	Interest on accounts		$20
		TOTAL	**$240**
		ANNUAL INCOME TOTAL	**$43,240**

KATE ALIBERTI

ANNUAL EXPENSES				
Real Estate Payments				
CitiBank	30-year fixed-rate loan, maturing in 2035 $185,000 to be paid in monthly installments of $1,139, split with husband			$6,840
	TOTAL	**$6,840**		
Property Taxes and Assesments				
Personal Property ("car tax")	$3.10 per $100 of assessed value			$316
Real Estate Tax	$1.20 per $100 of assessed value, Bill of $3,072, split with husband			$1,852
	TOTAL	**$2,168**		
Federal and State Income Taxes				
Federal Taxes	Married rate, 15%			$6,450
State Taxes	State of Virginia, based on tax table			$2,213
			TOTAL	**$8,663**
Other Loan Payments				
VISA (MBNA)	$200/month			$2,400
			TOTAL	**$2,400**
Insurance Premiums				
Geico	$400/month			$4,800
Erie Insurance (house)	$672, total bill split with husband			$336
			TOTAL	**$5,136**
Living Expenses				
Utilities	Total bill split with husband			$1,530
Food and Entertainment	Budgeted $150/week			$7,800
			TOTAL	**$9,330**
ANNUAL EXPENSES TOTAL	**$34,537**			

Business Resume

SAMANTHA BAUM

100 MAIN BROADWAY · RICHMOND VIRGINIA 23220
PHONE 804.555.5490 · EMAIL BAUMSR@VCU.EDU

WORK EXPERIENCE

Bloomingdale's 2003 to Present
New York, New York
Buyer, Women's Accessories

- An East-Coast-centered department store with departments in clothing, accessories, and home furnishings.
- Evaluate current fashion trends with respect to target customer's lifestyle, income, and preferences in order to select seasonal accessory inventory.
- Collaborate with designers and manufacturers to ensure that the selection exceeded customer's expectations.
- Received the Bloomingdale's "Innovator Award" 2005.
- Increased sales by 25 percent over a three-year period.

Arden B. 2000 to 2003
Richmond, Virginia
Store Manager

- A nationwide private-label boutique specializing in contemporary women's clothing and accessories.
- Responsible for all aspects of store operations including sales, personnel management, inventory control, and merchandise display.
- Increased the store revenue 23 percent through aggressive sales plan and expense control.
- Decreased employee turnover rate to 5 percent through selective hiring and team-building activities.
- Receive corporate recognition as the Southeastern region "Store of the Year" for outstanding customer and employee satisfaction ratings.

EDUCATION

Virginia Commonwealth University May 2003 Graduate
Richmond, Virginia
B.A. Fashion Merchandising

- Graduated Magna Cum Laude.

Personal Financial Statements

SAMANTHA BAUM

ASSETS			
Cash and Cash Equivalents			
Westfield Bank	Checking Acct. # 101300251		$2,420
Westfield Bank	Savings Acct. # 101300252		$7,200
		TOTAL	**$9,620**
Cash On Hand			
Westfield Bank	Cash at home		$450
		TOTAL	**$450**
Marketable Securities			
Whole Foods Market, Inc	50 shares on NasdaqGS		$3,000
Pfizer Inc.	50 shares on NYSE		$1,300
Netflix	50 shares on NasdaqGS		$2,675
		TOTAL	**$6,975**
Cash Value of Life Insurance			
Whole Life Policy	USAA Life Insurance-		$3,685
		TOTAL	**$3,685**
Accounts and Notes Recievable			
Unsecured Loan	Lisa Baum, mother; payable $100 monthly		$1,500
		TOTAL	**$1,500**
Personal Property			
2006 Dodge Neon			$9,000
Jewelry Collection			$1,000
Laptop Computer			$800
		TOTAL	**$9,800**
Other Assets			
Cash Gift from Grandmother			$5,000
		TOTAL	**$5,000**
		ASSETS TOTAL	**$37,030**

SAMANTHA BAUM

LIABILITIES AND NET WORTH			
Credit Cards and Revolving Credit Account			
USAA			$4,000
VISA (Wachovia)			$3,750
		TOTAL	**$7,750**
Loans Against Life Insurance Policy			
USAA	$4,000 against policy; 48 months at 5.8% interest		$2,500
		TOTAL	**$2,500**
Other Liabilities			
Father	Monthly payments,$200 for 12 months		$2,400
		TOTAL	**$2,400**
		LIABILITIES TOTAL	**$12,650**
		ASSETS TOTAL	**$37,030**
		LIABILITIES TOTAL	**($12,650)**
		NET WORTH TOTAL	**$24,380**
		LIABILITIES AND NET WORTH TOTAL	**$37,030**

SAMANTHA BAUM

ANNUAL INCOME

Gross Salary and Wages			
Bloomingdale's			$58,000
		TOTAL	**$58,000**
Dividends and Interest			
Westfield Savings	$7,200 at 5%		$360
Whole Foods Market, Inc	Dividends on 50 shares		$50
Pfizer Inc.	Dividends on 50 shares		$200
Netflix	Dividends on 50 shares		$150
		TOTAL	**$760**
		ANNUAL INCOME TOTAL	**$58,760**

SAMANTHA BAUM

ANNUAL EXPENSES

Real Estate Payments or Rent			
Clacahan Properties	30-year fixed-rate loan, maturing in 2035		$12,000
		TOTAL	**$12,000**
Federal and State Income Taxes			
IRS			$14,000
State			$2,500
		TOTAL	**$16,500**
Insurance Premiums			
State Farm	Car insurance, $250 per quarter		$1,000
		TOTAL	**$1,000**
Living Expenses			
Food and Necessities	Budgeted $270/week		$14,040
		TOTAL	**$14,040**
		ANNUAL EXPENSES TOTAL	**$43,540**

HSH Associates, (2007). Amoritization calculator. Retrieved April 25, 2007, from HSH Associates Financial Publishers Web site: http://www.hsh.com/calc-amort.htm

Kleinman, Rebecca (2007, March). Fall guides. *WWD Atlanta*, 6-7, 22-23.

Nordstrom, (2007). Dress shop. Retrieved April 7, 2007, from Nordstrom Web site: http://shop.nordstrom.com/C/2374331/0~2376776~2374327~2374331? mediumthumb-nail=Y&origin=leftnav&pbo=2374327

Saks Fifth Avenue, (2007). Classic, modern and contemporary apparel: Dresses. Retrieved April 7, 2007, from Saks Fifth Avenue Web site: http://www.saksfifthavenue.com/main/ProductArray.jsp?FOLDER%3C%3Efolder_id =2534374303668271&PRODUCT%3C%3Eprd_id=845524446153187&ASSORT-MENT%3C%3East_id=1408474395222441&bmUID=1175982848809&Special=V

US Census Bureau, (2006, August 15). American factfinder. Retrieved April 7, 2007, Web site: http://factfinder.census.gov/home/saff/main.html?_lang=en

Glossary

accountant—One who is skilled in the practice of accounting or who is in charge of public or private accounts. An accountant is responsible for reporting financial results, whether for a company or for an individual, in accordance with government and regulatory authority rules.

action plan—A simple list of all of the tasks that you need to carry out to achieve an objective.

amortized cost—The distribution of a single lump-sum cash flow into many smaller cash flow installments, as determined by an amortization schedule.

angel investor—Individual who provides capital for a business start-up, usually in exchange for ownership equity.

asset—Any item of economic value owned by an individual or corporation, especially that which could be converted to cash. Examples are cash, securities, accounts receivable, inventory, office equipment, real estate, a car, and other property. On a balance sheet, assets are equal to the sum of liabilities, common stock, preferred stock, and retained earnings.

atmospherics—Retail-store factors such as display design and fixtures, flooring, smell, sound level, store lighting and temperature, wall coverings, and other elements of store's ambience, which can be studied and controlled by a retailer to influence the consumer's buying mood.

balance sheet—A company's financial statement that reports the assets, liabilities, and net worth at a specific time.

bankruptcy—A legally declared inability or impairment of ability of an individual or organization to pay creditors.

below the line—In accounting, items in an account that are excluded from the account total, such as appropriations and extraordinary items that have no effect on the profit or loss in the current accounting period.

Better Business Bureau (BBB)—An organization based in the United States, Canada, and Puerto Rico. The BBB states its purpose is to act as a mutually trusted intermediary between consumers and businesses to resolve disputes, to facilitate communication, and to provide information on ethical business practices (www.bbb.org).

bookkeeper—One who systematically records a company's financial transactions.

brand promise—A guarantee made to the customer that addresses such issues as company vision, quality, value, and customer service.

branding—Branding allows a company to differentiate itself from the competition and, in the process, to bond with its customers to create loyalty. In this way, a position is created in the marketplace that is much more difficult for the competition to poach. A satisfied customer may leave, but a loyal one is much less likely to.

break-even analysis—The break-even point for a product is the point where total revenue received equals the total costs associated with the sale of the product (TR=TC). A break-even point is typically calculated in order for businesses to determine if it would be profitable to sell a proposed product, as opposed to attempting to modify an existing product instead so it can be made lucrative. Break-even analysis can also be used to analyze the potential profitability of an expenditure in a sales-based business.

business plan—A formal statement of a set of business goals, the reasons why they are believed attainable, and the plan for reaching those goals. It may also contain background information about the organization or team attempting to reach those goals.

capital—Wealth in the form of money or property, used or accumulated in a business by a person, partnership, or corporation.

capital equipment—A fixed asset with a useful life of a year or more that is not permanently attached to a building.

capital spending plan—A written document that details the spending on physical assets such as equipment, property, or a building. For accounting purposes, these items generally have a useful life to the business of more than one year.

cash flow—The amount of cash a company generates and uses during a period, calculated by adding non-cash charges (such as depreciation) to the net income after taxes. Cash flow can be used as an indication of a company's financial strength.

census—The process of obtaining information about every member of a population. The term is mostly used in connection with national "population and door-to-door censuses" (to be taken every 10 years as required by the U.S. Constitution).

census tract—A particular community defined for the purpose of taking a census. Usually these coincide with the limits of cities, towns, or other administrative areas, and several tracts commonly exist within a county.

Chamber of Commerce—An association of businesspeople to promote commercial and industrial interests in the community.

classifications—Also called categories, refers to a grouping or assortment of merchandise that customers view as interchangeable.

collateral—A security pledged for the repayment of a loan.

collusion—An agreement, usually secretive, that occurs between two or more persons to deceive, mislead, or defraud others of legal rights, or to obtain an objective forbidden by law. Collusion typical involves fraud or gaining an unfair advantage and can involve wage fixing, kickbacks, or misrepresenting the independence of the relationship between the colluding parties.

commission—A fee for services rendered based on a percentage of an amount received or collected or agreed to be paid (as distinguished from a salary).

commodity business—A business that competes primarily on the basis of price normally characterized by high asset intensity, significant capital expenditures, low profit margins, and intense competition. Examples of commodity businesses are airlines and textile companies.

common area fees—Mandated additional expenses added to the monthly rental fee or mortgage payment. These may be made up of such expenses as maintenance of common grounds (grass mowing and snow removal, for example); trash removal; repair of buildings and grounds; common marketing efforts; and others.

competition— Rivalry between two or more businesses striving for the same customer or market.

cost of goods sold—The total cost of buying raw materials and paying for all the factors that go into producing finished goods.

credit line—The maximum amount of credit to be extended to a customer. Also called *line of credit.*

credit rating—An estimate of the amount of credit that can be extended to a company or person without undue risk.

current assets—Accounts receivables, inventory, work in process, cash, and so forth that are constantly flowing in and out of a firm in the normal course of its business, as cash is converted into goods and then back into cash. In accounting, any asset expected to last or be in use for less than one year is considered a current asset.

current liabilities—Obligations such as deferred dividend, trade credit, and unpaid taxes, arising in the normal course of a business and due for payment within a year. Also called current debt.

debt-to-worth ratio—Denotes relationships of items within and between financial statements—for example, current ratio, quick ratio, inventory turnover ratio, and debt/net worth ratios.

deliverable—Report or item that must be completed and delivered under the terms of an agreement or contract.

demographics—Refers to selected population characteristics as used in government, marketing, or opinion research.

departments—Grouping of merchandise that can stand alone as a specialty store.

depreciation—A term used in accounting, economics, and finance with reference to the fact that assets with finite lives lose value over time.

destination shop—A store whose amenities, service, and products attract customers regardless of its location or even the cost of its merchandise.

direct competitor—A similar business that is most likely to attract the same target customer for the same type of merchandise.

entrepreneurship—The practice of starting new organizations or revitalizing mature organizations, particularly new businesses, generally in response to identified opportunities.

equity—In accounting: (1) Ownership interest or claim of a holder of common stock (ordinary shares) and some types of preferred stock (preference shares) of a firm. On a balance sheet, equity represents funds contributed by the owners (stockholders) plus retained earnings or minus the accumulated losses. (2) Net worth of a person or firm computed by subtracting total liabilities from the total assets.

executive summary—The section of a business plan that is a summary of the highlights of a business plan. It is also called the letter of intent.

FICA—The federal law that requires employers to withhold a portion of employee wages and pay them to the government trust fund that provides retirement benefits. An acronym for Federal Insurance Contributions Act. More commonly known as Social Security.

fixed cost—A cost that remains constant, regardless of any change in a company's activity. An example is a lease payment, which remains the same throughout the term of the lease.

foreclosure—The equitable proceeding in which a bank or other secured creditor sells or repossesses a parcel of real property (immovable property) due to the owner's failure to comply with an agreement between the lender and borrower, called a "mortgage" or "deed of trust."

franchise—A business method that involves the licensing of trademarks and methods of doing business, such as a chain store that shares a brand and central management, or an exclusive right to sell branded merchandise.

FUTA—Federal Unemployment Tax Act tax is a payroll or employment tax paid solely by the employer. While the FUTA tax is paid by the employer, it is based on each employee's wages or salary. Generally, the FUTA tax ends up being 0.8 percent of the first $7,000 per year of each employee's wages or salary. That means the employer's maximum cost for FUTA per year per employee is $56 ($7,000 x 0.008).

geographic footprint—The area in which the business wants to sell its product or services.

gross margin—A company's total sales revenue minus its cost of goods sold.

impulse items—Retail items known for their unplanned purchases and, therefore, kept near the checkout counters, such as candy, chocolate, magazines, novelties, and snacks.

income—The money that is received as a result of the normal business activities of an individual or a business.

income statement—A financial statement for companies that indicates how revenue (money received from the sale of products and services before expenses are taken out, also known as the "top line") is transformed into net income (the result after all revenues and expenses have been accounted for, also known as the "bottom line").

indirect competitor—A business that sells products or services that fill the same need as your business.

initial markup—Initial markup (IMU) is the difference between the initial retail selling price (at the time of receipt of the merchandise) and the cost.

insurance—The equitable transfer of the risk of a loss, from one entity to another, in exchange for a premium.

interest—The fee charged by a lender to a borrower for the use of borrowed money.

interest rate—A rate that is charged or paid for the use of money. An interest rate is often expressed as an annual percentage of the principal. It is calculated by dividing the amount of interest by the amount of principal.

Internet—Sometimes called the "Information Superhighway," the Internet is a worldwide, publicly accessible series of interconnected computer networks that transmit data by packet switching using the standard Internet Protocol (IP). It is a "network of networks" that consists of millions of smaller domestic, academic, business, and government networks, which together carry various information and services, such as electronic mail, online chat, file transfer, and the interlinked Web pages and other resources of the World Wide Web (WWW).

inventory—The merchandise selection that a business has available to sell.

inventory control—Maximizing turnover rate, stock-to-sales ratio and sales per square foot of a store's merchandise selection through careful planning and wise scheduling of such factors as when merchandise is received into store inventory and when markdowns are taken.

inventory management—Efficient control of product availability while maximizing profits or minimizing costs and meeting or exceeding customer expectations.

investor—A person or entity that commits money or capital in order to gain a financial return.

justifications—Also called assumptions, provide an explanation of the source and thought processes behind numbers in financial documents.

key items—Within each product category, those items that are deemed most important to the generation of sales.

key ratios—Financial performance measures used by business analysts in evaluation of the financial position and income of a firm. These include ratio of net income to average assets (called return on assets, or ROA), net income to equity (called return on equity, or ROE), net income to total investment (called return on investment, or ROI), and of net income to outstanding shares (called earnings per share, or EPS). Key ratios give a reliable indication of the competence of a management, especially when compared with the figures of previous periods or those of the competitors.

lease—Agreement in which one party gains a long-term rental agreement and the other party receives a form of secured long-term debt.

letter of intent—The section of a business plan that is a summary of the highlights of a business plan. It is also called the executive summary.

liability—A financial obligation, debt, claim, or potential loss.

line of credit—The maximum amount of credit to be extended to a customer. Also called *credit line*.

liquid asset—An asset that is easily and cheaply turned into cash—notably, cash itself and short-term securities.

logo—A graphical element (ideogram, symbol, emblem, icon, sign) that, together with its *logotype* (a uniquely set and arranged typeface) form a trademark or commercial brand. Typically, a logo's design is for immediate recognition, inspiring trust, admiration, loyalty, and an implied superiority.

long-term liabilities—Obligations payable in goods or services at a future period—more than 12 months away from today or the date of balance sheet. A firm must disclose its long-term liabilities in its balance sheet with their interest rates (or other charges) and date of maturity.

maintained markup—Markup after all reductions are taken. Reductions include markdowns, stock shortages, and employee and customer discounts.

markdown—A decrease in the selling price of an item.

market condition report—Also called a market *feasibility report*, the purpose of this research is to identify strategic opportunities in the focused market for a new business.

marketing plan—A written document that details the necessary actions to achieve one or more marketing objectives.

market weeks—A traditional purchase period when buyers begin to make selections for the next season.

media kit—Also called a *press kit*, this reference is often used in business environments. It is a prepackaged set of promotional materials of a person, company, or organization distributed to members of the media for promotional use.

mission statement—A statement that addresses what business the company is in, its primary purpose, strategies, and values.

net worth—For a company, total assets minus total liabilities. Net worth is an important determinant of the value of a company, considering it is composed primarily of all the money that has been invested since its inception as well as the retained earnings for the duration of its operation.

niche—A focused, targetable portion (subset) of a market sector.

non-cash expenses—Expenses that appear on the cash flow statement yet do not actually represent a real cash outflow.

overstocks—A quantity of merchandise for sale that is more than necessary or desirable.

owner's draw—Sums of money the proprietor or partners of a firm choose to take out of the cash flow of the business for personal use.

payroll services—Businesses that offer similar core services: calculating payroll and tax obligations for each employee, printing and delivering checks, and providing management reports. Payroll services can handle business for pay periods that run weekly, biweekly (every other week), semimonthly (twice a month), or monthly.

professional employer organization (PEO)—Sometimes used by small businesses who wish to have an outside company handle their human resource responsibilities. In these cases, the employees work for the small business, but a separate company manages all human resource responsibilities such as payroll; benefits; insurance; and federal issues, including taxes and workman's compensation.

personal financial statement (PFS)—Formal records of a person's financial activities. These statements provide an overview of a person's financial condition in both short and long term.

price points—Prices at which demand is relatively high.

prime loan rate—Interest rate charged by banks to their largest, most secure, and creditworthy customers on short-term loans. This rate is used as a guide for computing interest rates for other borrowers. Also called prime rate.

profit—The making of gain in business activity for the benefit of the owners of the business.

profit and loss plan—A set of limits that determines the maximum loss or gain a business can sustain.

pro forma—Assumed, forecasted, or informal information presented in advance of the actual or formal information. The common objective of a pro forma document is to give a fair idea of the cash outlay for a shipment or an anticipated occurrence. Pro forma financial statements give an idea of how the actual statement will look if the underlying assumptions hold true.

pro forma income statement—A financial statement constructed from projected amounts. A firm might construct a pro forma income statement based on projected revenues and costs for the following year.

psychographic data—The use of data to study and measure attitudes, values, lifestyles, and opinions for marketing purposes.

purchases—Product selections to be bought at wholesale for store inventory.

qualitative—Relating to or involving studies based on qualities.

quantitative—Relating to or involving studies based on quantities.

receipts—Merchandise that has been accepted into store inventory.

reciprocal of markup (also known as *cost compliment*)—This is the number left after subtracting the markup or planned markup for an item or category of goods from 100%. It represents the total cost percentage (or Cost of Goods) for the item, category, department, or even total store merchandise selection.

resume—A document containing a summary or listing of relevant job experience and education, usually for the purpose of obtaining an interview when seeking employment.

Retail Merchants Association—A group of businesses joined together to enhance the image and profitability of member companies through advocacy, information, and networking opportunities.

retainer—A fee charged in advance to retain the services of someone.

sales per square foot—Most commonly used for planning inventory purchases. Calculated by dividing the gross sales by the total square footage of the selling space. When measuring sales per square foot, keep in mind that selling space does not include the stockroom or any area where products are not displayed.

sales plan—List of specific steps that a business uses to determine how it will sell its products or services.

sales representative—A person employed to represent a business and to sell its merchandise (as to customers in a store or to customers who are visited).

SCORE—For Service Corps of Retired Executives, a nonprofit association dedicated to entrepreneur education and the formation, growth, and success of small businesses nationwide. SCORE is a resource partner with the U.S. Small Business Administration.

selling square footage—

1. The amount of square footage to be allocated for the display of merchandise to be sold.
2. A factor in the financial calculation of sales per square foot that indicates the amount of sales activity that results from only those parts of the store space that

are specifically allocated to the display of merchandise. For example (except in rare cases such as shoe stores), the stock room, office space, bathrooms, and the area behind the cash and wrap would not be a part of selling square footage. Other areas of the store may or may not be included; for example, if dressing rooms have mirrors, they would be part of selling square footage.

shrinkage control—Finding ways to positively influence the difference between book inventory and physical inventory due to counting or recording errors, or resulting from pilferage, spoilage, theft, or wastage.

simulation—An imitation of some real thing, state of affairs, or process.

Small Business Administration (SBA)—A United States government agency that provides support to small businesses.

stock keeping unit (SKU)—Identification numbers given to an item of merchandise. Typically, an SKU is associated with any purchasable item in a store or catalog. For example, a woman's blouse of a particular style and size

summary planning tools—Financial planning documents (pro forma statements such as the cash flow forecast, income statement, and balance sheet) that help to minimize risk of failure and maximize the probability of success. They crystallize all of the planning by anticipating the future results of the business.

SWOT Analysis—A strategic planning tool used to evaluate the Strengths, Weaknesses, Opportunities, and Threats involved in a project or in a business venture.

target customer—A group of clients and consumers that a business serves or wants to serve.

target market—The market segment to which a particular product is directed. It's often defined by age, gender, geography, and/or socioeconomic grouping.

vanilla box—Term used to describe a newly built store that has only four walls, a roof, and perhaps a bathroom; the rest of the interior design of the store is up to the tenant.

value added—Of or relating to the estimated value that is added to a product or material at each stage of its manufacture or distribution.

vendor—Anyone who provides goods or services to a company.

venture capitalist—Professional, outside investor that provides private equity capital to new businesses.

visual merchandising—The art of implementing effective design ideas to increase store traffic and sales volume.

Web site—A collection of Web pages, images, videos, or other digital assets that is hosted on one or several Web server(s), usually accessible via the Internet.

wholesale costs—The price that a store owner must pay to purchase product/goods from a wholesaler. The difference between the retail price and the wholesale cost represents the gross margin that a store will realize on the full retail price sale of an item.

working capital—Current assets minus current liabilities. Working capital measures how much in liquid assets a company has available to build its business.

Resources

Web Site Resources:

Commercial real estate sites: Use to find store locations and information.

www.loopnet.com
www.grubb-ellis.com
www.cushwake.com
www.craigslist.com

Government sites:

www.sba.gov: Small Business Administration main site
www.business.gov: General source for government help
www.eeoc.gov: Equal Employment Opportunity Commission, for help with labor laws
http://www.irs.gov/businesses/index.html: Internal Revenue Service for help with tax issues
http://www.firstgov : Government-specific search engine
http://www.google.com/ig/usgov: Google's government search engine
http://www.fedstats.gov: The government's gateway to federal statistics and statistical agencies.
http://www.bea.gov: U.S. Bureau of Economic Analysis has U.S. economic statistics.

Note: Much of the compiling of this information was done by Bettina J. Phifer, librarian, Virginia Commonwealth University.

General sites:

www.Bplans.com: Business plan resource site
www.hr.blr.com: Human resources site
www.careerplanner.com: Job description resource site
www.bizstats.com: Useful information about statistics that affect business planning
www.Hoovers.com: Company overviews, financials, executive bios, news, and competitor information

Consumer research:

http://factfinder.census.gov: Database from the U.S. Census Bureau with population, housing, economic, and geographic data
www.census.gov: Information as above

www.bls.gov: Bureau of labor statistics contains information about inflation and consumer spending; wages, earnings, and benefits; productivity; safety and health; international economic conditions; occupations; employment and unemployment; industries; and business costs.

http://www.harrisinteractive.com: Polling and public opinion data for information about consumers. Information provided by the Gallup Organization, Harris, and Roper Center for Public Opinion Research.

http://www.mediamark.com: Comprehensive demographic, lifestyle, product usage, and exposure to all forms of advertising media collected from more than 26,000 personal interviews with consumers

www.claritas.com: Demographic data, population estimates, maps, market segmentation

http://yawyl.claritas.com: Home of Prizm Market segmentation research

Trade Associations

http://www.cottoninc.com: Cotton, Inc. has information about apparel and home furnishings made with cotton. They also write the *Lifestyle Monitor*, an excellent source of information consumer attitudes and the fashion industry.

http://www.nrf.com: National Retail Federation

http://www.apparelandfootwear.org: American Apparel and Footwear Association, USA trade organization for U.S. apparel and footwear manufacturers, suppliers, and retailers

Trade Publications

Women's Wear Daily: For womenswear

Daily News Record: For menswear

Footwear News: For shoes

Accessories: For jewelry, hats, handbags, and other accessories

Home Fashion News: For home and gift items

Chain Store Age: Retail store information

Financial Ratios

Almanac of Business and Industrial Financial Ratios

Industry Norms and Key Business Ratios

RMA Annual Statement Studies

Hint for students: You can probably find these sourcebooks in your school library.

Books and Periodicals

Rand McNally Commercial Atlas and Marketing Guide: Provides marketing data in graphic format. Includes population, economic, and geographic data for more than 120,000 U.S. places.

Community Sourcebook of County Demographics

Community Sourcebook of Zip Code Demographics

Demographics USA: Demographic, economic, and commercial/ industrial data including Designated Market Area (DMA) rankings, Effective Buying Income (EBI), and Buying Power Index (BPI) by county and retail sales by store group and merchandise line

The Lifestyle Market Analyst: Correlates demographic characteristics with consumer buying behavior

The Wall Street Journal

New Strategist publications: Publishes volumes that synthesize statistical data from various government and private sources. Some of the titles are the following:

- *Americans and Their Homes:* Demographics of homeownership
- *American Incomes:* Demographics of who has money
- *The American Marketplace:* Details demographics and spending patterns in easy-to-read tables and graphs
- *American Men:* Who they are and how they live
- *American Women:* Who they are and how they live
- *The Baby Boom:* Americans born 1946 to 1964
- *Generation X:* Americans born 1965 to 1976
- *Older Americans:* A changing market
- *The Millennials:* Americans born 1977 to 1994
- *Who's Buying . . . ?* (Various topics, including apparel, home furnishings, entertainment—by race and for pets)

Bibliography

Abrams, Rhonda. *Business Plan in a Day*. Palo Alto, CA: The Planning Shop, 2005.

Arrington, Deidra. *Planning Inventory the Right Way* (Presentation). Available at www.aetaexpo.com/html/seminar_content.html, education tab, August 2007.

Bell, Judith, and Kate Ternus. *Silent Selling*. 3rd ed. New York: Fairchild Books, 2005.

Berry, Tim. "Writing an Executive Summary." Bplans.com. Available at http://articles.bplans.com/index.php/business-articles/writing-a-business-plan/writing-an-executive-summary

Blanchard, Ken, and Sheldon Bowles. *Raving Fans: A Revolutionary Approach to Customer Service*, New York: William Morrow, 2003.

Bohlinger, Maryanne Smith. *Merchandise Buying*. 5th ed. New York: Fairchild Books, 2001.

Brandenburger, Adam, and Barry J. Nalebuff. *Co-Opetition*. New York: Doubleday Business, 1996.

Buchanan, R., and C. Gilles. "Value Managed Relationship: The Key to Customer Retention and Profitability." *European Management Journal* 8, no. 4 (1990).

Dion, Jim, and Ted Topping. *Start and Run a Retail Business*. Washington, D.C.: Self-Counsel Press, 1998.

Empire State Development. Your Business: *A Guide to Owning and Operating a Small Business in New York State*. Brochure.

Falk, Edgar A. *1001 Ideas to Create Retail Excitement*. Rev. ed. New York: Penguin Group, 2003.

Georgia State University Small Business Development Center. *Entrepreneur Risk Assessment Quiz for Starting Your Own Business*. June 2007. Available at http://www2.gsu.edu/~wwwsbp/newwebsite/entrepre.html.

Granger, Michele, and Tina Sterling. *Fashion Entrepreneurship: Retail Business Planning*. New York: Fairchild Books, 2003.

Grant, Jeff. *The Budget Guide to Retail Store Planning and Design*. Cincinnati, OH: ST Media Group, 1995.

Guthrie, Karen M., and Cynthia W. Pierce. *Perry's Department Store: A Buying Simulation for Junior, Men's Wear, Children's Wear, and Home Fashions/Giftware* 2nd ed. New York: Fairchild Books, 2003.

Harroch, Richard D. *Small Business Kit for Dummies*. Foster City, CA: IDG Books, 1998.

———. "Types of Commercial Real Estate Leases." January 2005. Available at http://www.allbusiness.com/operations/facilities-real-estate-office-leasing/2611-1.html.

Henshell, John. "Guide to Setting Employee Pay Levels and Salaries." Available at http://www.business.com/directory/human_resources/compensation_and_benefits/salary/#guide.

Holland, Phil. *Business Expansion and Handling Problems*. MyOwnBusiness.org. Available at http://www.myownbusiness.org/s12/#topten.

HSH Associates. (2007). Amortization calculator. Retrieved April 25, 2007, from HSH Associates Financial Publishers Web site: http://www.hsh.com/calc-amort.htm

Isis, Inc. *Empowering Business through Technology*. Brochure. 2002.

Jacobs, Robin. *Types of Employees in the Workplace*. Portland, OR: Office for Students with Disabilities, Portland Community College, 2003. Available at http://spot.pcc.edu/~rjacobs/career/types_of_employees.htm.

King, Jan. *Business Plans to Game Plans: A Practical System for Turning Strategies into Action*. Rev. ed. New York: John Wiley & Sons, 2003.

Kleinman, Rebecca. (2007, March). Fall guides. *WWD Atlanta*, 6–7, 22–23.

Kuebler, Ginny L. *Building Your Future in Self-Employment*. New York: GLK Management Consulting.

———. *Let's Write Your Business Plan: A Step-by-Step Workbook for All Start-Up Small Businesses*. 7th ed. New York: GLK Management Consulting, 2001.

Lopez, Michael J. *Retail Store Planning and Design Manual*. 2nd ed. New York: John Wiley & Sons, Inc., 1995.

Main Street Beaufort, USA. *Wrong Assumptions about Destination Businesses*. Available at http://www.downtownbeaufort.com/uparticle/Wrong%20Assumptions%20About%20Destination%20Businesses.pdf.

McKeever, Mike P. *How to Write a Business Plan*. 6th ed. Berkeley, CA: Nolo, 2002.

Miller, Julie. *Start Your Own Clothing Store: Your Step-by-Step Guide to Success*. Canada: Entrepreneur Media, 2003.

Minassian, Mark. *The Different Types of Bank Loans*. Available at http://biztaxlaw.about.com/od/financingyourbusiness/a/types_of_loans.htm.

Nelson, Bob, Peter Economy, and Ken Blanchard. *Managing for Dummies*. 2nd. ed. January 2003. Available at http://www.dummies.com/WileyCDA/DummiesArticle/id-933,subcat-BUSINESS.html?print=true.

Nordstrom. (2007). Dress shop. Retrieved April 7, 2007, from Nordstrom Web site: http://shop.nordstrom.com/C/2374331/0~2376776~2374327~2374331? mediumthumbnail=Y&origin=leftnav&pbo=2374327.

Raphael, Todd. "Happiness May Be Overrated." Brief article. *Workforce*. (May 2002.)

———. Richmond District Office. *Small Business Resource Guide* 2003. Brochure.

Saks Fifth Avenue. (2007). Classic, modern and contemporary apparel: Dresses. Retrieved April 7, 2007, from Saks Fifth Avenue Web site: http://www.saksfifthavenue.com/main/ProductArray.jsp?FOLDER%3C%3Efolder_id=2534374303668271&PRODUCT%3C%3Eprd_id=845524446153187&ASSORTMENT%3C%3East_id=1408474395222441&bmUID=1175982848809&Special=V.

Service Corps of Retired Executives. *Personal Financial Statement Form*. Score.org. Available at http://www.score.org/downloads/Personal%20Financial%20Statement1.xls.

———. *Social Security Handbook*. January 2008. Available at http://www.ssa.gov/OP_Home/handbook.

Thomas, Charles. "Top tips for Dealing with Business Disasters." Entrepreneur.com.

Available at http://www.entrepreneur.com/management/operations/article78212.html
Timm, Paul R., Ph.D. *50 Powerful Ideas You Can Use to Keep Your Customers.* 3rd ed. Franklin Lakes, NJ: Career Press, 2002.
U.S. Census Bureau. (2006, August 15). American factfinder. Retrieved April 7, 2007, Web site: http://factfinder.census.gov/home/saff/main.html?_lang=en.
U.S. Small Business Administration. *Managing Employee Benefits.* SBA Publication PM-3.

Credits

Step 1

1.1 Courtesy of *Heidi Story*, designer & owner of Heidi Story, Richmond, Virginia

Step 2

2.1 Courtesy of Casey Longyear & Marshe Wyche, owners of Rumors, Richmond, Virginia
2.2 Courtesy of Heidi Story, designer & owner of *Heidi Story*, Richmond, Virginia
2.4 Courtesy of Randy Rollins, president & owner © Martha's Mixture, Ltd., 2008 and Thalhimer, Inc. Realty, Richmond, Virginia
2.5-2.7 Design by Jennifer Hamilton, Interior Designer, Professor Virginia Commonwealth University Department of Interior Design

Step 3

3.1 Courtesy of Fairchild Publications, Inc.

Step 5

5.1 Courtesy of Libby Sykes & Deborah Bouchen, owners *Pink*, Richmond, Virginia
5.2 Courtesy of Lisa McSherry, owner *Lex's of Carytown*, Richmond, Virginia
5.3 Courtesy of Heidi Story, designer & owner of *Heidi Story*, Richmond, Virginia
5.4 Courtesy of Heather Teachey-Lindquist, owner of *Que Bella*, Richmond, Virginia
5.5 Courtesy of Casey Longyear & Marshe Wyche, owners of *Rumors*, Richmond, Virginia

Step 6

6.1-6.6 Courtesy of Fairchild Publications, Inc.
6.7 © Erin Fitzsimmons
6.8 Courtesy of Casey Longyear & Marshe Wyche, owners of *Rumors*, Richmond, Virginia

Step 7

7.1 Created by Michael A. Sisti
7.2-7.7 Copy & Design by Michael A. Sisti, rose illustration by Krutika Kotval

Step 8

8.1-8.3 Courtesy of Fairchild Publications, Inc.

Step 9

9.1-9.4 Courtesy of Fairchild Publications, Inc.
9.5 Design by Jennifer Hamilton, Interior Designer, Professor Virginia Commonwealth University Department of Interior Design
9.6 Courtesy of Casey Longyear & Marshe Wyche, owners of *Rumors*, Richmond, Virginia

Conclusion

C.1 © Hans Neleman/Taxi/Getty Images
C.2 Courtesy of Fairchild Publications, Inc.
C.3 Veer
C.4 Courtesy of Fairchild Publications, Inc.
C.5 Alamy

Index